AF686352

WAR MACHINE

First edition published in 2026 by Uitgeverij
An imprint of punctum books, Earth, Milky Way
https://www.punctumbooks.com

ISBN-13: 978-1-68571-321-8 (paperbound)
ISBN-13: 978-1-68571-322-5 (PDF)
ISBN-13: 978-1-68571-323-2 (EPUB)
DOI: 10.53288/0564.1.00
LCCN: 2026939754
Library of Congress Cataloging Data is available from the Library of Congress

Editing: Vincent W.J. van Gerven Oei & Eileen A. Fradenburg Joy
Book design: Hatim Eujayl
Cover design: Vincent W.J. van Gerven Oei

p. punctumbooks

spontaneous acts of scholarly combustion

RICHARD A. CARTER

War Machine

Contents

Notes on the Text	7
Plates	35
Sources	119

Notes on the Text

War Machine is speculative sounding of the myriad entanglements of technology, ecology, discourse, politics, and conflict shaping the planetary environment. In this context, objects, activities, and environments cannot be readily separated from one another, or their varied affects distilled into clear lines of causality and analytical insight, challenging efforts at narrating their characteristics through words alone. It is on this basis that *War Machine* has been created not as a conventional academic text, but in the mode of generative, more-than-textual criticism and poetry, driven by "esoteric" algorithmic processes. The aim in doing so is to depict juxtapositions of intra-acting phenomena whose frictions, contrasts, tensions, and ambiguities can be registered better through emergent material performances, both in the processes of composition and on the page itself, rather than through the crisp, clipped certainties of conventionally written scholarship and its attended vectors of argument. In short, *War Machine* deploys experimental formal techniques in an attempt to *enfold* and *perform*, rather than simply describe, an uncertain fusion of competing discourses and activities. The subsequent stream of outputs and interpretive potentials necessarily precludes the ascendency of any "definitive" critical narration and resists any straightforward integration with current academic canons of worldly diagnosis. Nevertheless, it is forwarded here that speculative practices such as this can facilitate the creation of different stories, and especially different *kinds* of stories, concerning current challenges. Speculative work is well placed to model adventurous, forward-facing approaches, across both thought and practice, that refuses inaction when dealing with extremely challenging scenarios. It can acknowledge head-on varied problematics and their attended uncertainties and doubts, through sheer material necessity, while still contributing to the aesthetic, affective, and conceptual foundations from which alternative modes of worldly knowing and being can be developed and justified. The outcome of such work is to instantiate, once more, how the conjunction of art and criticism can respond meaningfully to the fierce difficulties of the present moment, especially when their immediate powers of rectification are often minimal, and when the harms engaged are too widespread and too far embedded to ever fully mitigate.

***

The core premise of this text involves staging a digitally simulated wargame that stands in for the multitude of political, rhetorical, and economic battles that are presently raging across the face of a profoundly endangered planet. Featured are two starkly competing depictions of the Earth: 1) as a body facing extreme threats to its climatic and ecological stability; 2) as a body of valuable resources to be extracted, managed, policed, and contested as part of commercial enterprise and geopolitical strategy. The algorithmic operations behind this wargame are employed as a generative procedure to assemble a flowing textual approximation of these contrasting worldviews, imaginaries, and rhetorics as they intersect, reframe, and compete with one another, their weight of meaning and influence ever-shifting and being driven by the very energies and materials they attempt to account for, albeit in very different ways. The antagonistic discourse yielded by this process is invariably noisy, fraught, and uncertain, manifesting across a multiplicity of registers and contexts. It oscillates between supporting and undermining the statements made, while nevertheless remaining enframed within a global system of unrestrained extractivism. It was the desire to map this sense of saturated frictions, collisions, and contrasts—the sheer *noise* and *intensity* of contemporary material-discursive exchange—that initially guided the composition of *War Machine.*

***

The actual wargames enacted by this text are held between two teams of "drones" that seek to establish strategic control over the game space. The drones themselves are represented as abstract symbols, modelled loosely on US military "Common Warfighting Symbology" unit markers, and they maneuver across a terrain consisting of military Tactical Pilotage Charts (TPCs) depicting environments marked on a "Red List" of endangered ecosystems maintained by the International Union for the Conservation of Nature (IUCN). This Red List has been devised to act as a common resource informing decision makers about current ecological threats, their management, and conservational requirements, as well as facilitating public awareness-raising (IUCN 2023).

The rules of the game are that each drone is assigned an arbitrary strength value and a grid position on "friendly" terrain near allied units. The strength values of each drone radiate outwards, diminishing in proportion to distance, across a gridded "influence map," and these values can either combine with those of neighboring allies or be negated entirely by the presence of those belonging to the opposition. The terrain itself can also pose some measure of resistance, in eroding the tactical "reach" of a given drone. It is this undulating influence map that furnishes each drone with the necessary spatial and tactical awareness to operate. The drones employ a simple form of AI in order to advance across the terrain toward opposing units, attempting to project and combine their radiating zones of influence so as to numerically cancel out or even wholly envelope and eradicate those of the opposition. If a drone finds itself in a space where the opposing influences surrounding it are greater than its own, it is eliminated from the field, although it endeavors to avoid this fate, if possible.

It is this global influence map that is depicted in the visual half of each poem, with the current location of each drone at a particular timestep being plotted along with a symbolic representation of their tactical status (maneuvering, searching, engaging). Also featured are their intersecting zones of influence, which are demarcated as a numerical field of positive and negative integers. This field subsequently highlights areas of collective strength and weakness between the opposing forces, as well as neutral areas of no overall control.

It is the sequential unfolding of each wargame—evolving in ways similar to that of a cellular automaton—that furnishes a generative mechanism for assembling a granular form of cut-up text. Each grid point of the global influence map under the control of a particular drone faction is used to select a co-located character belonging to an assigned text. As described above, these texts are often rhetorically opposed along numerous vectors: either reporting on the serious dangers facing global habitability or conducting detailed market analyses, with environmental breakdown cast as a risk to supply chains and investment returns. In some cases, the texts involved are of the same kind and effectively compete (even if inadvertently) as part of a noisy information environment that can only be processed within finite envelopes of critical attention.

After a few discrete timesteps of the wargame have passed, the grid field is sampled, and the current mixture of textual characters is then printed opposite to the game visualisation, manifesting as a shifting cloud of contrasting expression, characterized by an uncertain layout, spacing, and fragmentary wording.

* * *

There are several ways that the functional, visual, and textual elements of *War Machine* might be parsed, whether in terms of how to characterize the individual components of each poem-diptych, or to consider what they all might ultimately mean at scales and durations beyond the immediate themes of the text as a whole. The chief power of creative-critical exploration is, in part, its capacity to denaturalize what we understand an *analytical* response to a body of varied themes, concepts, and concerns might resemble and to facilitate a perspective on them that centers not on the formulation of a diagnostic account—with the pretence of there being clearly separable chains of effect that can be readily understood and (notionally) resolved—but on sounding their generative possibilities for meaning and insight, particularly along the material and affective vectors often sidelined in academic writing, and which often better characterize our experiences and reactions towards a rapidly changing world.

These rarefied aspirations aside, however, it should also be acknowledged that convention and constraint can rarely be foregone entirely in published work, and considerable caution should always be sounded against any claim, implicit or otherwise, that a text or art object is readily accessible, even self-explanatory, for all audiences across all contexts. In the aforementioned spirit of using friction and ambiguity generatively in order to tell a variety of possible tales, it is therefore worth outlining a few notes concerning what are the more evident themes engaged by *War Machine*—those which often served as its direct inspiration—and to provide some entryways into how they might be considered.

* * *

War Machine emerged partly in reflection of the long history of wargaming itself, as a tool for simulating and predicting battlefield contingencies. Modern professional wargaming can be traced back to the Prussian *Kriegsspiel* in the early 19th Century (Peterson 2023), and continues to be a prominent feature in the context of military training and planning, whether manifest in the form of physical maps and models or computerized emulations of such (Brynen 2020; Reddie et al. 2018). The subsequent application of "wargaming," often more as a planning concept than a rules-based model, is also apparent in various other domains that are predicated on managing and responding to evolving challenges, contingencies, and threats, such as political strategizing and business forecasting (Kurtz 2003).

While ecological factors do often play a role in contemporary military wargaming, in terms of accounting for its logistical and operational impacts, they have occasionally featured in the context of security response planning for a changing world, with the opposing "forces" being the deteriorating planetary environment. In one instance (Boulton 2023), the climatic-ecological "hyperobject"—in Morton's sense (2013) of a hyperdistributed set of relations exceeding all possible observational frames—is parsed as a centrifugal "hyperthreat" which needs to be tackled across all fronts. This particular characterization of environmental breakdown as a quasi-military challenge, amenable to the logics and rhetorics of security operations and strategic "threat postures," is an uncomfortable one, given how military activities are an essential part of the forces perpetuating the dangers at play (Parkinson and Cottrell 2021). This invites subsequent reflection on whether a continuation of these outlooks could in any way contribute to the transformative shift in political and economic consciousness required (a point forwarded by Warner and Boas 2019). The overall posture of privileged nations toward the wider planet can certainly be characterized as that of a war machine, seeking to obtain the total subjugation of Earthly resources, and to maximize the logistical efficiencies by which these can be extracted and applied, irrespective of the costs borne by the humans and animals living in their vicinity, or the stability of the global environment. It is here that the concept of wargaming commercial scenarios and threats to business growth have far more immediate resonance, and the blunt image of geopolitically framed, commercially underpinned battles over fragile (but potentially resource-rich) ecological spaces was one of the darker points of inspiration behind *War Machine.*

The evident counter to this bleak perspective is in restating a broader vision of gameplay as a generative process for highlighting and simulating structural issues in the world—or, at least, fostering an attitude that emphasizes the possibility of concrete action over its presupposed futility (Wilkinson 2016). We can cite here the growing body of recent table tops titles that engage ecological themes. Some titles, such as *PARKS* (2019), *TRAILS* (2021) or *Fjords* (2022) focus on exploration. Other titles emphasise the detailed simulation or even transformation of the Earthly environment, such as *Photosynthesis* (2017), *Ecology* (2019), *Canopy* (2021), and *Daybreak* (2023). In contrast, *digital* games tend to frame natural spaces in a largely exploratory, narrative, or ambient mode, rather than as a system to be engaged, e.g., *Flower* (2009), *Firewatch* (2016), *Walden, a game* (2017), *Eastshade* (2019), *Beyond Blue* (2020), and *Naiad* (2024). A few more strongly educational or activist titles do exist, however, that acknowledge the presence and agency of the more-than-human world, e.g. the *Shelter* series (2013–2021), *Lumino City* (2014), *Never Alone* (2016), and *Lichenia* (2019). In some cases, however, such as *Terra Nil* (2023) or *Planet Crafter* (2024), the processes of tackling ecological harms are presented as a perfect reclamation of a polluted or desolate space—an effective return to Eden that erases materially embedded histories and harms.

It is interesting to consider that across all these varied instances, the modeling of player agency in relation to the environment stands in marked contrast to the apocalyptic visions of Earthly collapse that are often found in many ecologically charged settings across the wider media landscape, where the principal interest is on combative human struggles in desperate circumstances, rather than on mitigating the harms ensnaring all. While *War Machine* is a self-playing game, and one that depicts the forces driving present challenges, it has nevertheless been made in reflection of the value of games as offering a structure for studying and creatively reimagining different environmental scenarios, inviting onlookers to step outside the immediacies of events driven by certain assumptions of action, outside the "rules of play," and to keep in view the continuing potentials to enact different modes of material engagement, with different possible outcomes arising.

* * *

The technical architectures used in *War Machine* for modeling the forces of conflict and competition were based on the use of computerized influence maps to guide the behavior of individual units in a coherent way. This technique is found within some simple digital wargaming simulations, and is closely aligned with the modeling of spatial AI behaviors and navigation pathfinding within computer games more generally (Mark 2019). The influence propagation algorithm is known specifically as Dijkstra's Algorithm, which was developed originally by mathematician

Edsger Dijkstra for finding the shortest paths between nodes in a weighted graph. A gridded game space can be treated as just such a graph, and by overlapping versions of this space with different weighted features (such as allied and enemy units, or terrain obstacles) it is possible to define objective nodes and trace a path toward them that is responsive to factors beyond distance alone, emulating "intelligent" decision-making.

This detail is worth highlighting because it gives cause for reflecting on the necessary narrowness of all algorithmic operations, and the wider consequences these can generate, despite contemporary excitement around certain forms of "AI" as finally overcoming these limitations. That is, as a simple tool to emulate intelligent pathfinding, Dijkstra's algorithm can only operate on weighted factors that are explicitly fed into it—even if the behaviors generated are not programmed directly as such. In *War Machine,* the game terrain, already a synthetic abstraction, features only relatively weakly in the calculus for the deployed units, and this is a conscious expression of how the overwhelming majority of algorithmic logics governing contemporary life give no regard to their wider environmental contexts, beyond those immediately necessary for executing their central taskings. Although this might appear a wholly irrelevant concern from the pragmatic perspective of software engineering, just as the contemporary built world still instantiates political and economic logics that are predicated on abundant, readily available energy (chiefly from fossil fuels), the logics of the digital world model and assume all actions will take place within broadly stable environmental contexts that will not exceed certain thresholds of change. While this might be a necessary assumption when creating algorithms that function in abstractly digital contexts (e.g., calculate the optimal connections between points in a fixed body of data), if the process is such that it is highly energy intensive, in order to be achievable within reasonable timescales, then the material implications beyond the computational become significant, both economically and environmentally (Caballar 2024). We might cite Clark's (2012) concept of "scale-effects" as an illustration here, in which actions and gestures at one scale can generate impacts at scales far beyond the perceptual and cognitive frames in which they are otherwise "rationally" justified. Computer science and software engineering certainly do place immense value on both speed and efficiency, particularly when facing technological constraints, but as shown by the recent, ballooning expansion of generative AI systems, these pressures can fall away rapidly if the excess hardware capability exists and the economic return is sufficient.

In this regard, there is currently great concern that generative AI systems, in terms of their text and image synthesizing operations, the training demands of creating new models, and the entire supply chain of materials and labor involved in their operation, are placing wholly unsustainable burdens on the Earthly environment (see Crawford 2024; Valdivia 2024), undercutting oft-cited claims that they are critical tools for modeling and managing the very threats being faced (Brevini 2020). These burdens manifest chiefly in the form of the already vast power draw of large data centers and the water consumed in their cooling (see, e.g., Ren 2023; Monserrate 2022), as well as risks that upcoming generations of consumer-grade, device-level AI activity will lead to more intensive computing operations on average. Proponents of the new systems seemingly pay little attention to these concerns, and speak of a wholesale revolution that will impact every facet of human life, across all contexts, overriding national borders, social disparities, and geographies (Leaver and Srdarov 2023). The result, it is forwarded, will be an age of hyperabundance that will upend conventional economic realities and transform how human beings live in the world at large.

Leaving aside how realistic these predictions or aspirations may ultimately be, it is worth reflecting on their presumption of a deep digital future that will remain untroubled by Earthly events. Prevailing digital imaginaries tacitly hold that the incredible demands of the latest AI systems, which can thrive only in the most rarefied ecological niches we might furnish, can be sustainably met at a global scale both now and into the far future. The notion that digital infrastructures are rendered intensely vulnerable precisely due to their seemingly unrestrained requirements in energy and materials is never broached. Futuristic imaginaries of an AI-transformed human world roundly ignore the growing contingencies of an ecologically deteriorating planet, with all the conflicts and catastrophes this engenders, and the serious risks these will pose to the globally distributed supply chains on which all digital technologies critically rely. In this regard, *War Machine* offers a performative enactment of the self-devouring nature of the digital environment and the sprawling economic ecologies on which it relies. Its constituent parts are not only in active competition with one another—with all the wastage and duplicated effort this implies—but also scour the resource-rich, often highly vulnerable terrains that are critical for their continued operation.

* * *

Another domain in which the implications of AI systems specifically are looming large is in the context of popular and scholarly debates around automated decision-making architectures, including autonomous weapon systems, and their erosion of the perceptual, conceptual, and legal parameters in which conflicts are defined, conducted, and adjudicated. Armed drones esepcially have long been totemic within these discussions. Even though they stand as being guided by human pilots, the latter's judgment and autonomy, operating as part of a highly integrated and automated surveillance, intelligence, targeting infrastructure, has often fallen under scrutiny (Emery 2020). Richardson (2024) has noted how contemporary technoscientific warfare often surpasses the perceptual and conceptual capacities of the human entities within them, even though the latter are entangled within widely distributed, technocultural platforms of perception, with sensory data and its interpretation shifting from an indexical to an operationalized set of relations. It is in this context that algorithmically crystallized biases and blind spots can contribute to patterns of observation and analysis that materialize the very activities that are "expected" to be seen, driving incidents such as "friendly fire" or misdirected strikes against purely civilian targets, or, in more recent cases, a rapid erosion of critical concern regarding the latter (Emery 2020; McKernan and Davies 2024), raising vital questions regarding where legal accountability ultimately lies.

One final area of widespread concern around the very latest modes of generative AI especially is its applicability for information warfare, whether by manufacturing outright disinformation (Goldstein et al. 2024) or by driving bot activities that pollute, destabilize, and misdirect online discourse such as to break down social trust and cohesion (Hajli et al. 2021). Even if explicitly political activity is still an emerging problem, the capacity for generative AI to accelerate the now unceasing thrum of online "noise" (the "dark media" fabrics of spam, malware, clickbait, and advertorials that erode information retrieval) is well discussed in the popular media, and it is this rendering of contemporary digital life as unpleasant, uncertain, and (at points) unusable that constitutes its chief "threat" (see, e.g., Heikkilä 2023 for a typical journalistic illustration of these concerns).

It is on the subject of generativity, however, that *War Machine* seeks to model an alternative way of thinking through what generative AI technologies and their outputs can potentially represent. Just as the core premise of creative gameplay, as outlined above, is to invite a variety of approaches and modes of seeing toward challenging scenarios, the process of turning the movements of AI conflict into a form of poetry models a vision in which these systems are no longer bound by some inherent destiny toward certain trajectories of use but are shown as the product of conscious choices within contexts that can nevertheless be modified. Richardson's (2024) discussions concerning artistic media resonate here, in noting the power of their material reimagining of what it means to sense and represent the world in an era of intersecting technological, martial, and ecological crises. Speculative gestures and creative redeployments of media, form, and convention, which lie at the very premise of this particular text, invite opportunities to reflect on and even enact conjunctions of aesthetics and worlds that do not adhere to dominant forms of knowledge-making and media imagination (9). That is, artistic reworkings of established conjunctions of materials and energies make space for other ways of knowing, imagining, and intervening in the world, establishing communicative relationships that trouble existing paradigms, highlight otherwise excluded potentials, and offer avenues for the narration of different stories concerning the capacities of worldly becoming (117).

A useful contextualizing reference regarding the characteristics of *War Machine* is that of so-called "esoteric" programming languages, or "esolangs." These are schemas of computer coding that are deliberately playful, puzzle-like, and ultimately strange compared to well established languages (Temkin 2017b). Instances here include Piet, a visual language designed to resemble abstract artworks (Temkin 2017a); Hexagony, in which the code consists of single character commands arranged along a hexagonal geometry (Temkin 2018); in:verse, which maps a table of words onto shader commands, enabling a written poem to generate an image (Temkin 2020a); or Cree# and 文言/Wenyan, which, for the former, adopts the Cree syllabic alphabet and storytelling traditions as a coding system (Temkin 2021), while the latter adheres to the tonal and grammatical conventions of classical Chinese literature (Temkin 2020b), utilizing traditional Chinese characters exclusively in its composition (both examples being rarities in computing languages that overwhelmingly employ Latin characters and English grammar).

There is an inherent playfulness to these languages, in denaturing established processes of instructing computers, and exploring ideas and concepts beyond the bald pragmatics of programming efficiency. They show us how the vernacular of code is reshaping cultural registers for expressing contemporary experiences and how code also embeds cultural assumptions concerning what "counts" as effective instructional practices.

War Machine is not itself an esolang, in that it employs conventional languages and runtime environments (a variation of Java), but it does offer what might be characterized as a form of esoteric programming, in which established processes for achieving a particular outcome are replaced with deliberately strange, circumlocutory operations that express concepts beyond the ends to which they are directed. Such endeavors have underpinned several of my previous projects, all of which generate poetic outcomes from unusual conjunctions of computing machinery and processes, meditating on what it means for the latter to apprehend and give account of more-than-human phenomena and environments. *War Machine* has been created in a similar vein, turning the bleak calculus of military wargaming and algorithmic combat into an apparatus for assembling reflexively critical concrete poetry, an "alternative AI" modality for generating text. As with esolangs, which Temkin (2023) characterizes as "explicitly refusing practicality and clarity," the point here is not to wholly replace established paradigms but to challenge their apparent stability and self-evidence, to suggest an excess of technical and expressive potential that can be engaged through the subversions of speculative practice, whereby creative insight and resistance are always possible even within the most rigidly overdetermined technical contexts.

***

It is here that an answer might be developed concerning the status of works such as *War Machine* at the present moment. Given the deeply challenging state of the world, it can be asked to what meaningful end creative-critical work can represent and resist the forces at play? Even accepting the above explanations regarding the formal and conceptual inspirations behind this text, the critique remains that such efforts, particularly at the scale conducted here, are unlikely, if ever, to draw the attention of the key actors involved, or have much immediate hope of even beginning to redress the scale of the ongoing harms.

In other fields of scholarly analyses and activist reflection, the nature and scale of global challenges is captured or reflected through a whole array of critical labels, such as "polycrises" or "permacrises" (Turnball 2021); "the Anthropocene/Anglocene/Capitolocene/Plantationocene/Thermocene/Trumpocene" (and other variations: Mentz 2020, 57 counts up to twenty-four); "eco-nihilism" (Mukherjee 2016); "necrocapitalism" (Banerjee 2018); or, more bluntly, "disaster capitalism" (Klein 2008).

All these varied labels address their own particular areas of interest across different historical durations, but they are all broadly diagnosing an Earthly context that, outside the reassuringly thoughtful registers of critical analyses, can be affectively appalling for those who would seek a better future for human and more-than-human alike. In an age characterized by the deleterious intersections of rapid climatic and ecological breakdown at a global scale; an increasingly polluted and unnavigable information environment; an aggressive proliferation of generative AI; growing algorithmic warfare and highly automated weaponry; and all set against a backdrop of zero-sum competition for land, resources, and zones of influence, it would appear straightforward to claim that human capacities to absorb and mitigate such challenges are now thoroughly overwhelmed, with the critical, aesthetic, and emotional registers necessary to conceptualize and tackle them becoming saturated. From this bleak perspective, what is left is effectively *noise:* the traces of disintegrating critical purchase, diffusing emotional resolve, and a breakdown in political efficacy and meaningful agency in the face of unrelenting, overwhelming challenges, along with the varying degrees of despair and indifference this might precipitate. We might readily see such dark visions in the fragmentary character of the poems generated by *War Machine.*

Nevertheless, the riposte to such despondent narratives is straightforward enough to voice, even if thoroughly daunting to implement: what else to do? To endlessly emphasize the worst-case interpretations can be as destructive as any of the forces that are so profoundly damaging the world at present—to effectively declare that there is nothing to be done, and to reiterate the popular, if potentially simplistic interpretation of Fredric Jameson's dictum that "it is easier to imagine the end of the world than the end of Capitalism" (2005, 199).

Leaving aside the perennial debates raised concerning the role and value of art in society at large, one resonant response to the issues described above, as they relate to this current project, can be found in the practices of so-called "permacomputing." A loose play on the much older concept of "permaculture," permacomputing recognizes the profound environmental impacts generated by computing machinery and infrastructures, inclusive of their manufacture, maintenance, and disposal, and ultimately strives to model alternatives that shift away from the maximalist approaches of commercial deployments (Mansoux et al. 2023). Amongst those practitioners that identify as being invested in permacomputing, the goal is to enact different digital activities and artifacts that emphasize a significantly reduced footprint in energy and materials, embracing the generative productivity of concrete *constraint*, and ultimately aiming to develop approaches and attitudes that recognize and attempt resilience in the face of unrelenting ecological and economic degradation in the deep future (Mansoux et al. 2023). Such efforts also encompass digital creative works that adopt certain principles along a similar vector, and which critique the techno-economic conditions that facilitate the development of resource-intensive computing systems.

War Machine is deliberately quite simple, compared to what it could have been if it had used the latest AI architectures to drive the behavior of its constituent units and the text generated. While it is therefore unrepresentative of the technologies being utilized currently for synthesizing media or even simulating battlefield contingencies, the effect is a relatively minimalist set of operations that bear out a model of computing activity in which constraint forms a key part of the overall gesture being enacted. Here, technological functioning and the surfaces of expression made available combine to enact a different narrative concerning the contemporary environment. Permacomputing strives for just this kind of modality, and while *War Machine* is perhaps not a direct instance, it nevertheless has been created in a similar spirit. The latter extends into the very rationale for pursuing such practices in the first example. The core premise of permacomputing forwards that computing technologies will continue to prevail even in the teeth of a dark and uncertain future and so rejects the apocalyptic narratives of total global collapse that undermine serious thinking about what possibilities can be found to facilitate human and more-than-human flourishing. It is ultimately a refusal of despair and a refusal of the inaction this can all too easily justify. Digital art, conducted in the guise of both contesting current paradigms, while also exploring their possibilities and alternatives, is one manifestation of this. *War Machine* has been created expressly with this thinking in mind, articulating current woes head-on, but also modeling a creative and critical response to such that, it is hoped, resists simplistic narratives of defeatism and finds opportunities amidst the fearsome complexities and challenges.

It is the speculative nature of this response that forms one final aspect of the generative rational behind *War Machine,* in considering what specifically critical interventions might resemble at a time when conventional academic responses appear hard-pressed to register the full breadth of wickedly intersecting and rapidly accelerating technical, political, and ecological dynamics. Overall, the consciously unconventional gesture at the heart of *War Machine* is an attempt at articulating a multiplicity of intra-acting phenomena that are effectively irreducible to any clear casual vectors, narratives, or "solutions" to the critical problems they pose—to conduct a messy sounding of a kaleidoscopic environment whose incongruities are often confounding, unsettling, and sometimes absurd. In trying to represent these aspects, the goal is to bear out the matrix of challenging play, possibility, and uncertainty that such unruliness invites—to emphasise opportunities to tell myriad transient narratives that do not resolve into delimited analyses, maintaining an openness toward the curation of contingent resonances and vignettes that become kernels of always provisional insights or meaning, or indeed, by virtue of their strangeness, inspire further queries. These generative material and critical exchanges both contract and unfold the complexities of the present moment, and ultimately, in the face of their sometimes frightening implications, try to step back into the flux from which all art and storytelling arises, with human mark-making being predicted on a necessary kinship with our more-than-human registers of being-in-the-world, which both enable and influence what gets made and what gets told.

To rearticulate this point, we can acknowledge that even within the contexts of technologies and practices that seem inimical to such, there always remain opportunities for modes of reading and modes of action that recognize how, at the scales and durations of Earthly existence, meaning is established by, and propagates through our entanglements with more-than-human agencies. Art and storytelling do not simply furnish passive depictions of our current conditions, but always manifest as concrete interventions within them, carrying significance beyond human understandings alone. The pursuit of these practices represent another kind of a dialogue with the Earth, very different to that conducted by military-industrial technoscience or even conventional scholarship, and which has the particular power of fostering a continuing openness to alternative modes of knowing and being that can otherwise be elided by prevailing technical and analytical orthodoxies. This does of course require a willingness to actually conduct such work, rather than only calling for such, and while individual efforts do not inherently translate into a fundamentally changed world, they are nevertheless a vital contributor to the conditions in which such endeavours can be deemed possible at all, fostering new imaginaries and new meanings in the face of forbidding futures.

* * *

Each diptych in *War Machine* can be explored in any order desired by the reader. Below each map is a short caption that states the name of the IUCN ecosystem it depicts, along with the specific TPC chart reference. The charts themselves were sourced from the open access Perry-Castañeda Library Map Collection at the University of Texas at Austin. Many of these charts are 1980s or 1990s vintage, and sometimes depict places, names, and even ecosystems that no longer exist.

* * *

Banerjee, Subhabrata Bobby. 2008. "Necrocapitalism." *Organization Studies* 29, no. 12: 1541–63. DOI: 10.1177/0170840607096386.

Boulton, Elizabeth. 2023. "Climate Security Wargaming: Taking on the Hyperthreat in Southeast Asia." *Climate Change & (In)security Project.* https://cciproject.uk/resourcesblog/climate-security-wargamingtaking-on-the-hyperthreat-in-southeast-asia.

Brevini, Benedetta. 2020. "Black Boxes, Not Green: Mythologizing Artificial Intelligence and Omitting the Environment." *Big Data & Society* 7, no. 2. DOI: 10.1177/2053951720935141.

Brynen, Rex. 2020. "Virtual Paradox: How Digital War Has Reinvigorated Analogue Wargaming." *Digital War* 1: 138–43. DOI: 10.1057/s42984-020-00004-z.

Caballar, Rina Diane. 2024. "We Need to Decarbonize Software: The Way We Write Software Has Unappreciated Environmental Impacts." *IEEE Spectrum* 61, no. 4: 26–31. DOI: 10.1109/MSPEC.2024.10491388.

Clark, Timothy. 2012. "Derangements of Scale." In *Telemorphosis: Theory in the Era of Climate Change,* Vol. 1, edited by Tom Cohen. Open Humanities Press. DOI: 10.3998/ohp.10539563.0001.001.

Crawford, Kate. 2024. "Generative AI's Environmental Costs Are Soaring—and Mostly Secret." *Nature* 626, no. 693. DOI: 10.1038/d41586-024-00478-x.

Emery, Joshua R. 2020. "Probabilities Towards Death: Bugsplat, Algorithmic Assassinations, and Ethical Due Care." *Critical Military Studies* 8, no. 2: 179–97. DOI: 10.1080/23337486.2020.1809251.

Goldstein, Josh A., John Chao, Scott Grossman, Alex Stamos, and Michael Tomz. 2023. "How Persuasive Is AI-Generated Propaganda?" *PNAS Nexus* 3, no. 2. DOI: 10.1093/pnasnexus/pgae034.

Hajli, Nick, Umar Saeed, Maryam Tajvidi, and Farshad Shirazi. 2021. "Social Bots and the Spread of Disinformation in Social Media: The Challenges of Artificial Intelligence." *British Journal of Management* 33, no. 3: 1238–53. DOI: 10.1111/1467-8551.12554.

Heikkilä, Melissa. 2023. "We Are Hurtling Toward a Glitchy, Spammy, Scammy, AI-Powered Internet." *MIT Technology Review,* April 4. https://www.technologyreview.com/2023/04/04/1070938/we-are-hurtling-toward-a-glitchy-spammy-scammy-ai-powered-internet/.

International Union for Conservation of Nature (IUCN). 2023. "How the Red List Is Used." *IUCN Red List.* https://www.iucnredlist.org/about/uses.

Jameson, Fredric. 2005. *Archaeologies of the Future: The Desire Called Utopia and Other Science Fictions.* Verso.

Klein, Naomi. 2008. *The Shock Doctrine: The Rise of Disaster Capitalism.* Penguin Books.

Kurtz, James. 2003. "'Business Wargaming': Simulations Guide Crucial Strategy Decisions." *Strategy & Leadership* 31, no. 6: 12–21. DOI: 10.1108/10878570310505550.

Leaver, Tama, and Stefanie Srdarov. 2023. "ChatGPT Isn't Magic: The Hype and Hypocrisy of Generative Artificial Intelligence (AI) Rhetoric." *M/C Journal* 26, no. 5. DOI: 10.5204/mcj.3004.

Mansoux, Aymeric, Brendan Howell, Dušan Barok, and Ville-Matias Heikkilä. 2023. "Permacomputing Aesthetics: Potential and Limits of Constraints in Computational Art, Design and Culture." *Ninth Computing within Limits*. DOI: 10.21428/bf6fb269.6690fc2e.

Mark, Dave. 2019. "Modular Tactical Influence Maps." In *Game AI Pro 360: Guide to Tactics and Strategy*, edited by Steve Rabin. CRC Press. https://www.taylorfrancis.com/chapters/edit/10.1201/9780429054969-9/modular-tactical-influence-maps-dave-mark.

McKernan, Bethan., and Harry Davies. 2024. "The machine did it coldly': Israel Used AI to Identify 37,000 Hamas Targets." *The Guardian*, April 3. https://www.theguardian.com/world/2024/apr/03/israel-gaza-ai-database-hamas-airstrikes.

Mentz, Steve. 2020. *Break Up the Anthropocene*. University of Minnesota Press.

Monserrate, Steven G. 2022. "The Cloud Is Material: On the Environmental Impacts of Computation and Data Storage." *MIT Case Studies in Social and Ethical Responsibilities of Computing*, January 27. DOI: 10.21428/2c646de5.031d4553.

Morton, Timothy. 2013. *Hyperobjects: Philosophy and Ecology after the End of the World*. University of Minnesota Press.

Mukherjee, Upamanyu Pablo. 2016. "Para-Religions of Climate Change: Humanity, Eco-Nihilism, Apocalypse." In *A Cultural History of Climate Change*, edited by Tom Bristow and Thomas Ford. Routledge.

Parkinson, Stuart, and Linsey Cottrell. 2021. *Under the Radar: The Carbon Footprint of Europe's Military Sectors*. European United Left, SGR, the Conflict and Environment Observatory. https://www.sgr.org.uk/publications/under-radar-carbon-footprint-europe-s-military-sectors.

Peterson, Jon. 2023. "A Game Out of All Proportions: How a Hobby Miniaturized War." In *Zones of Control: Perspectives on Wargaming,* edited by Pat Harrigan and Matthew Kirschenbaum. MIT Press.

Reddie, Andrew, Benjamin L. Goldblum, Kiran Lakkaraju, Jason Reinhardt, Michael Nacht, and Lina Epifanovskaya. 2018. "Next-Generation Wargames." *Science* 362, no. 6421: 1362–64. DOI: 10.1126/science.aav2135.

Ren, Shaolei. 2023. "How Much Water Does AI Consume? The Public Deserves to Know." *OECD.AI,* November 30. https://oecd.ai/en/wonk/how-much-water-does-ai-consume.

Richardson, Michael. 2024. *Nonhuman Witnessing: War, Data, Ecology after the End of the World.* Duke University Press.

Temkin, Daniel. 2017a. "Interview with David Morgan-Mar." *Esoteric.Codes,* February 10. https://esoteric.codes/blog/david-morgan-mar.

———. 2017b. "Language without Code: Intentionally Unusable, Uncomputable, or Conceptual Programming Languages." *Journal of Science and Technology of the Arts* 9, no. 3: 83–91. DOI: 10.7559/citarj.v9i3.432.

———. 2018. "Interview with Martin Ender." *Esoteric.Codes,* May 15. https://esoteric.codes/blog/martin-ender.

———. 2020a. "in:verse." *Esoteric.Codes,* June 30. https://esoteric.codes/blog/inverse.

———. 2020b. "Wenyan-lang." *Esoteric.Codes,* December 12. https://esoteric.codes/blog/wenyan-lang.

———. 2021. "Interview with Jon Corbett." *Esoteric.Codes,* March 30. https://esoteric.codes/blog/jon-corbett.

———. 2023. "The Less Humble Programmer." *Digital Humanities Quarterly* 17, no. 2. DOI: 10.63744/yqfhrnggfdbw.

Turnbull, Neil. 2021. "Permacrisis: What It Means and Why It's Word of the Year for 2022." *The Conversation,* November 11. https://theconversation.com/permacrisis-what-it-means-and-why-its-word-of-the-year-for-2022-194306.

Valdivia, Ana. 2024. "The Supply Chain Capitalism of AI: A Call to (Re)think Algorithmic Harms and Resistance through Environmental Lens." *Information, Communication & Society* 28, no. 12: 2118–34. DOI: 10.1080/1369118X.2024.2420021.

Warner, Jeroen, and Ingrid Boas. 2019. "Securitization of Climate Change: How Invoking Global Dangers for Instrumental Ends Can Backfire." *Environment and Planning C: Politics and Space* 37, no. 8: 1471–88. DOI: 10.1177/2399654419834018.

Wilkinson, Phil. 2016. "A Brief History of Serious Games." In *Entertainment Computing and Serious Games,* edited by Ralf Dörner, Stefan Göbel, Michael Kickmeier-Rust, Maic Masuch, and Katharina Zweig. Springer. DOI: 10.1007/978-3-319-46152-6_2.

PLATES

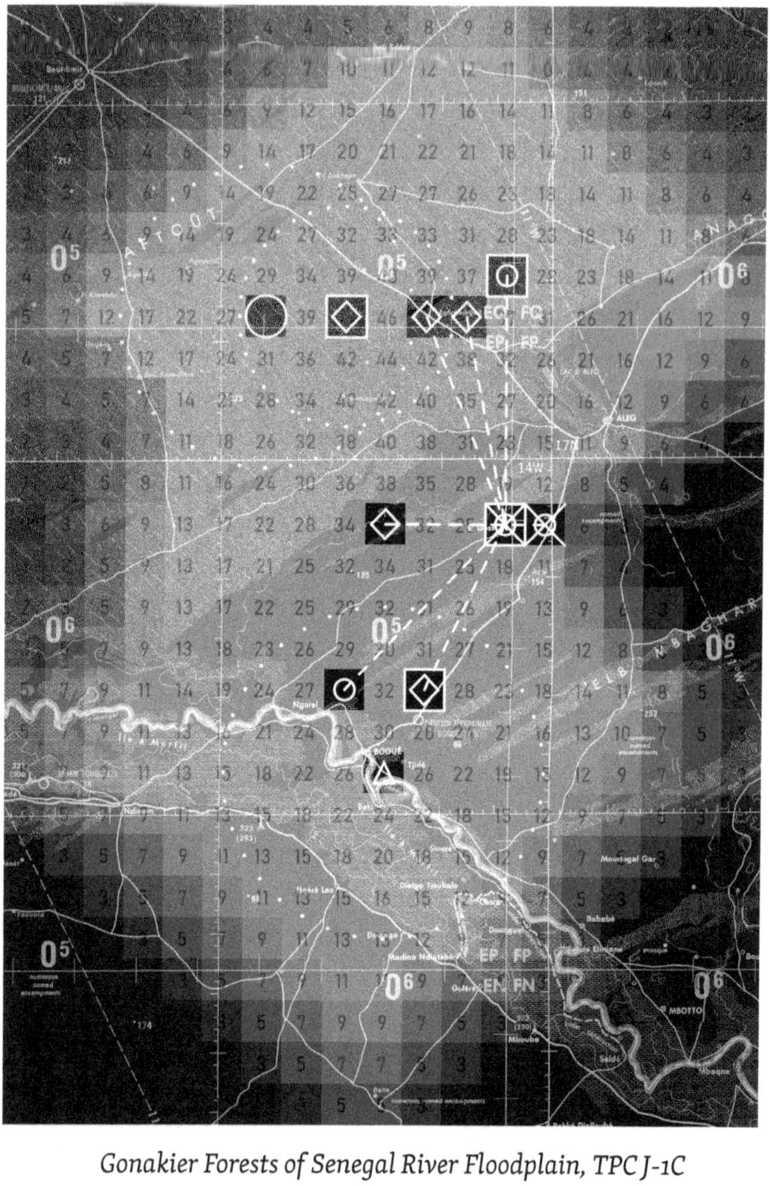

Gonakier Forests of Senegal River Floodplain, TPC J-1C

d however incre
ng climate risks
ill inevitably lea
d to more climate r
elated losses and d
amages therefore th
e adaptation gap re
port agr also focus
es on loss and dama
ge to support parti
es in the negotiati
ons following the d
cision at the twe
ty seventh sessio
of the conference
f the parties to th
e united nations fr
amework convention
on climate change c
op in sharm el shei
kh to establish a l
ss and damage fun
nd funding arra
nts for vulne
eveloping c
global te
and cli

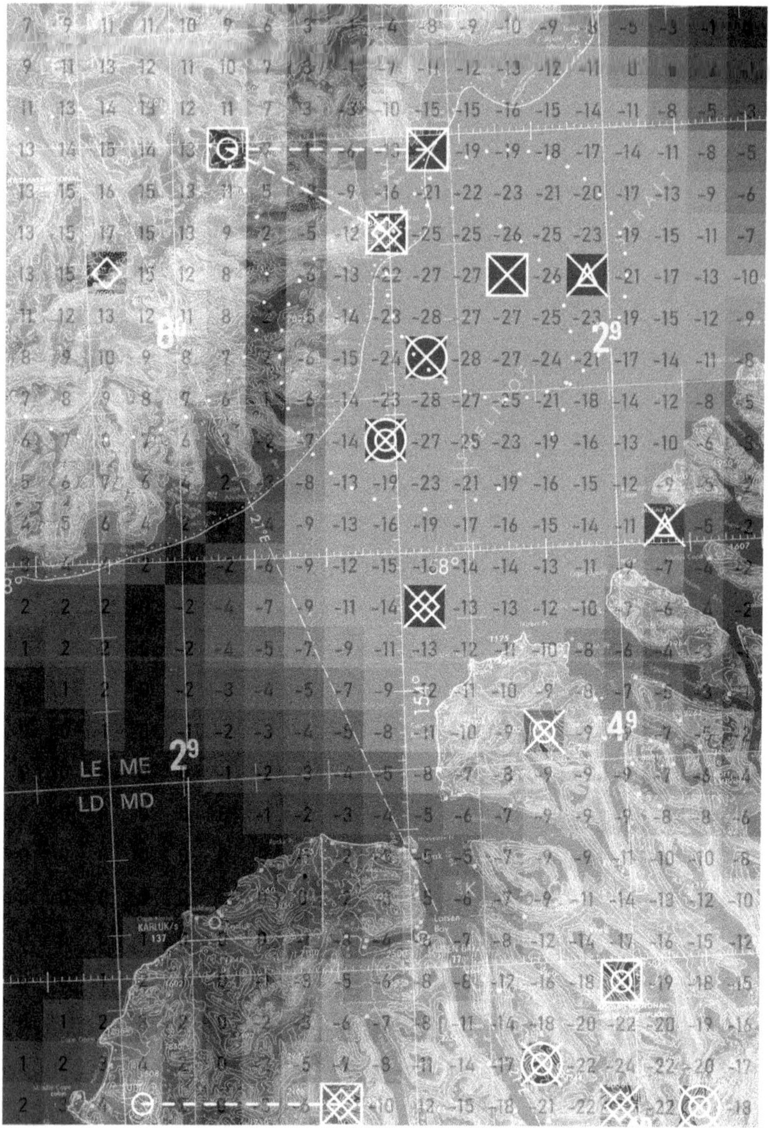

Giant Kelp Forests, TPC D-11D

on and f WELL OVER
dation iYEAR SO FAR
e prodes FROM ECUAD
projectsFELL AND IM
r inpe ROM ALL MAJO
deter dNG PARTNERS
and foWN SHARPLY I
tion alIRST TWO MON
ter defTHIS YEAR PA
lerts iY RAIN SLOWI
rest dRTIONS HEAVY
erts iLMOST DAILY I
est d ING OPERATION
rts GABON AND CONG
res ILE OPERATORS I
t d MEROON AND CENT
i AFRICAN REPUBLI
a E CURRENTLY EXP
CING A DRY PER
A RESULT OF
N PRODUCTION
RE DOWN AND
t TO THE FORE
gra CHALLENGING B
mazo BY CHINA PICK
concl DEMAND IN MI
ew re AST MARKETS I

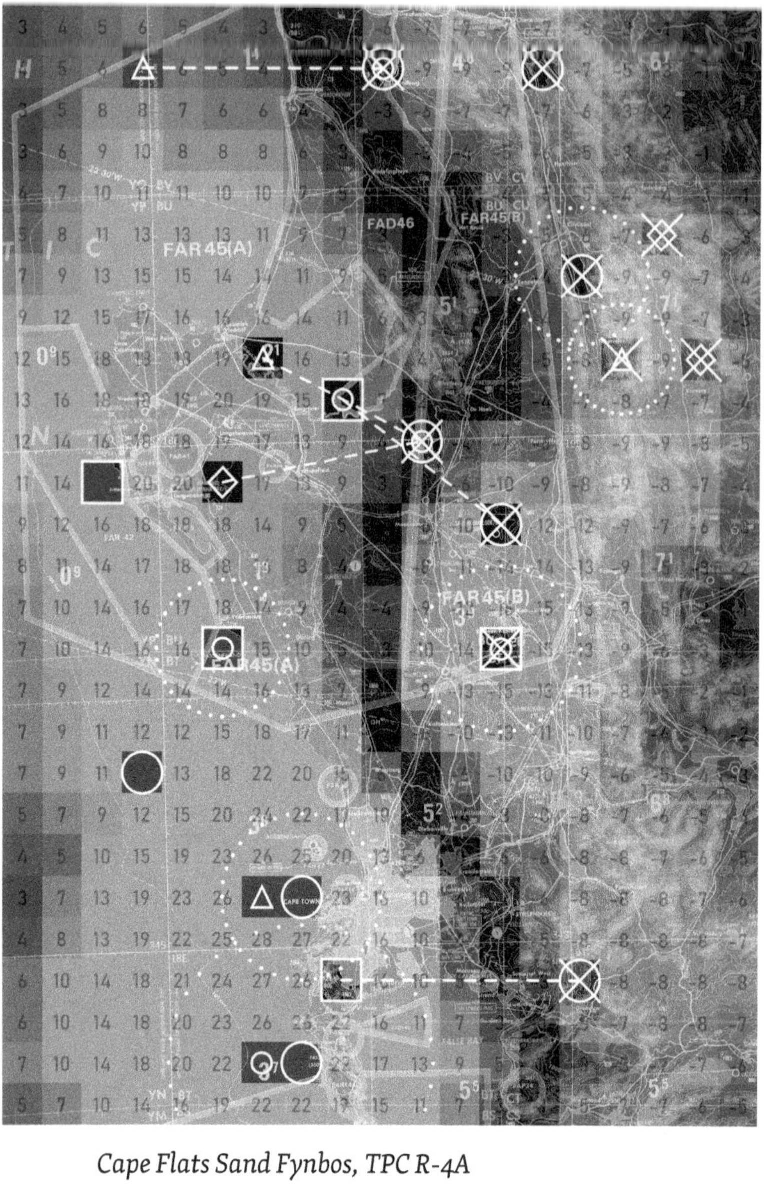

Cape Flats Sand Fynbos, TPC R-4A

agreed TRADE UP
hat this RD WAVE O
gas remo CAPACIT
should bCONVERSIO
how the ukEL PLANTS
net zero asT DELAYS
embers viewPANDEMIC
s topic are OVERHA
cant there i REFINE
her group t THIS CO
t once repH A SHARP
ve of the HINESE OI
tion and EXPORTS A
inted witEAVAL OF R
s of measADE FLOWS
red to reIN RECORD
ro recoveOR THE IND
majority YEAR WHI
ly membersINERY CAP
agreed or IONS OF
hat steps t D BY OU
the governme GROWTH
elp the econ PRODUC
over should bG TREN
gned to help UCTS M
e net zero whePEAT
ing their ratESS IN

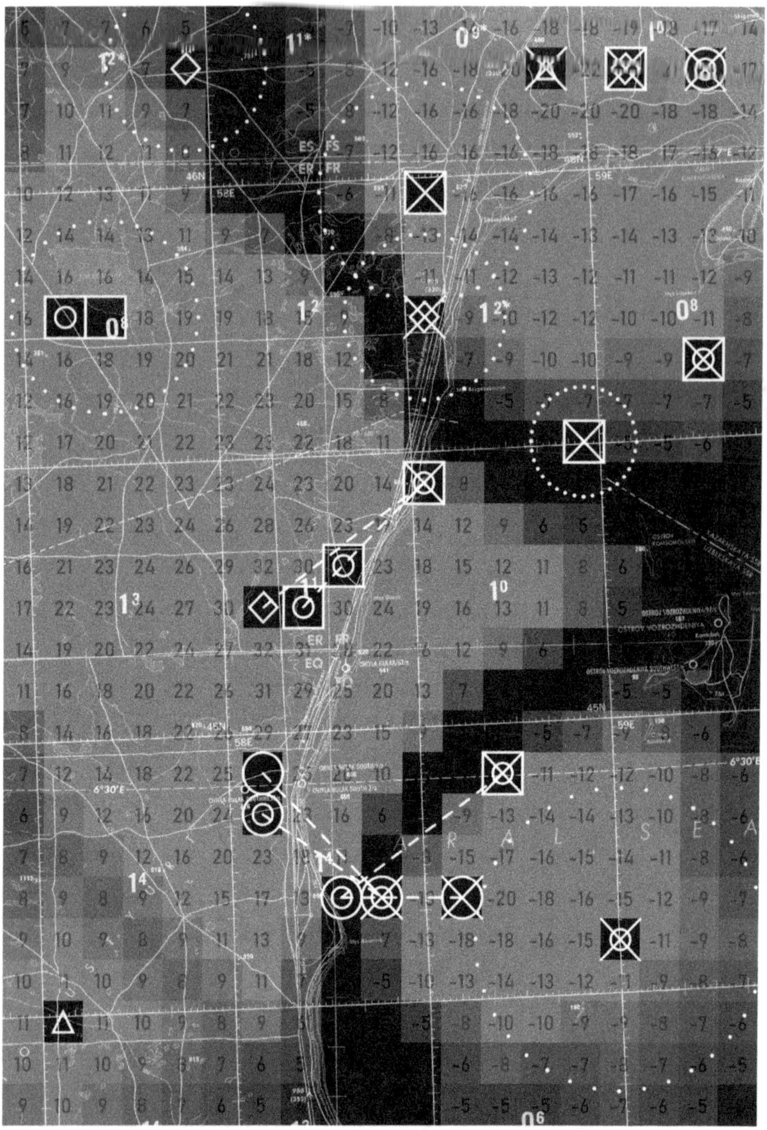

Aral Sea, TPC F-5A

m is nS INCLUDING O
e than AND COAL SUP
to aveD DEMAND RENE
orst iENERGY TECHNO
mate brELECTRICITY
world neENERGY EFFI
tors thosCESS TO EN
inspire a D SIDE MA
a modern ND MUCH M
n that cha ITS WOR
way the worl OCATES
if we are one WILL
he countries thatAB
ise to this global
challenge future ge
nerations could S M
sper and flourS ASS
e have the cNTRIES
ow to help THIS PUB
world to ND ANY MA
ct the in HEREIN A
ental panT PREJUDIC
ate changSTATUS OF
clear thIGNTY OVER
emissionsITORY TO T
ut by by TATION OF
on the rigNAL FRONT

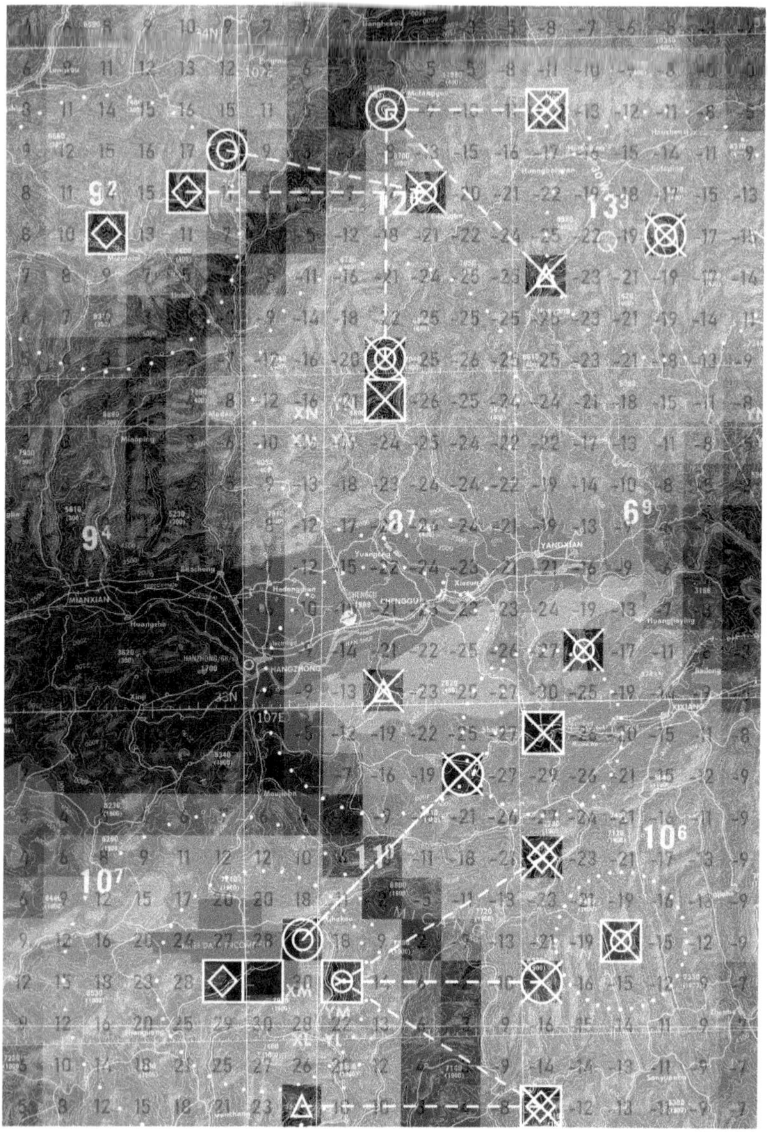

Great Lakes Alvar, TPC F-18A

nnovators LEUM EXPO
and buildTRIES JUL
ansform ouARTICLE T
and socie FOR THE
to bring IN OIL M
icity trLIGHTS F
d heatingCLE CRUD
into the s MOVEMENT
we need tY MARKETS
cally incrNOMY WORL
wable poweND WORLD
ure that c PRODUCT
s reap theD REFINE
by creati NS TANKER
ds of newRUDE AND R
cure jobsODUCTS TRA
rs in cleCIAL STOCK
ies we neS BALANCE
ate proof AND DEMAN
factorie SCLAIMER
ces makingNALYSIS
m in winterR INFORM
l in the suNFORMATI
ch will cut IN THE
and our bill MARKET
ed to protect MR IS
re and enhan IONAL

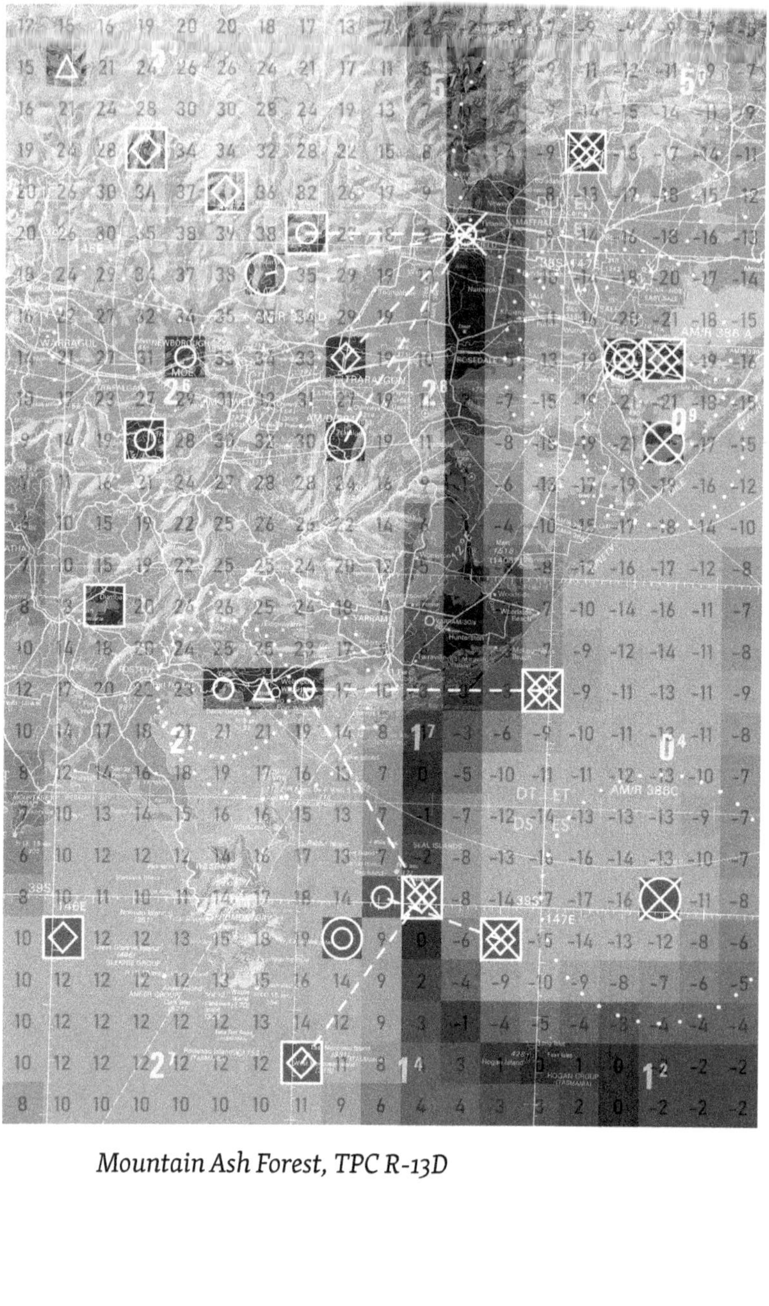

Mountain Ash Forest, TPC R-13D

in sweden sF CLIMAT
nd republic SIX CO
iye united HE MIDD
united stateH AFRIC
uropean commTENTS P
also particiKHALED
n the work oIA GETT
ea iea assoc CLIMAT
countries arATED TR
 brazil chinJECTION
 india indonBOVE PR
nya morocco S IMPLI
 singapore sAGRICUL
rica thailanD SECUR
ne abstract THE NAT
ugural editiMENT IN
he critical NTRIES
s market re THIS RE
vides a ma ALLY WHE
e on the iED AGAINS
 market teOUND OF O
and policyURES FOR
f the crit OCCO IS
rals sectorERIENCIN
an an initiNTED DRO
ng of the eme g YIN
picture for thr SIS

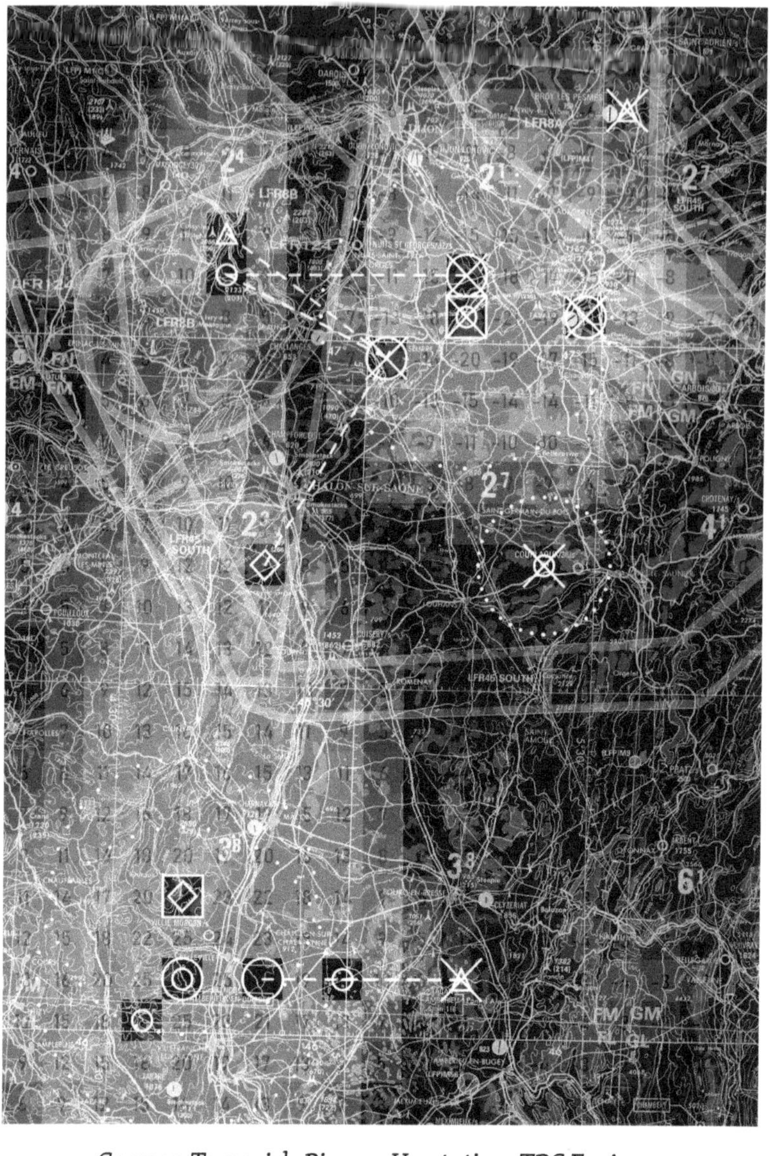

German Tamarisk-Pioneer Vegetation, TPC F-2A

nal FOR HUMA
s theUSINESS IH
s two TRADING NA
osals tE INSTITUT
dations AN RIGHTS A
emes incESS LTD LIM
sparencyGUARANTEE I
lity anATED IN ENGL
aking e WALES REGIS
municat FICE ADDRES
gement iERCIAL ROAD
l actionRNE EAST S
s and incE UK COM
yments conCHARITY
nd taxes adEGISTE
recommenda TENT
ssed by the AR
y for the fC I
and wordinTAR
h recommendS TH
me are detaNT AN
ase see chapORK C
ten recommeLIMATE
s that receiATION A
t support wESILIEN
transition RABILI
ero should LIMAT

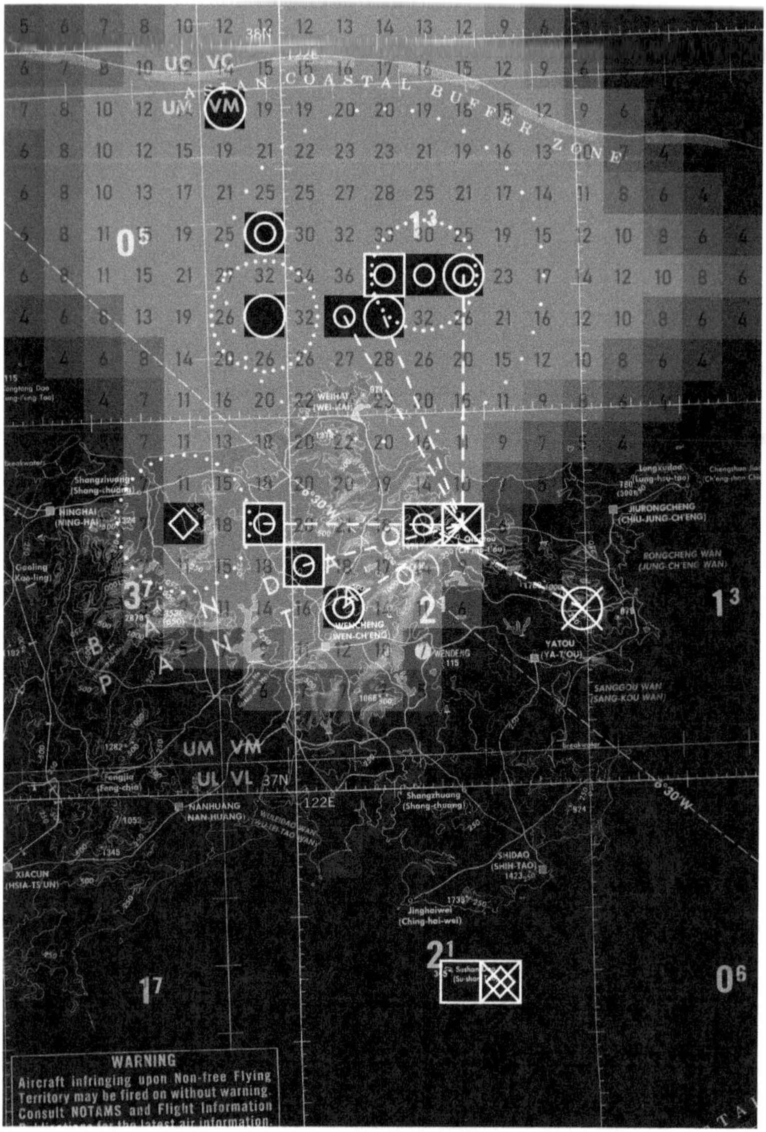

Tidal Flats of the Yellow Sea, TPC G-10A

```
grounds politic
ommitments as su
this collection b
ngs together a mul
itude of disciplina
ry conversations co
ncerned with art an
d aesthetics that a
re emerging around
 he anthropocene t
  is drawing toge
 r artists curat
 scientists thBY
 s and activi ENE
  ddress the O DIS
   l reforma OAL
     hu man  N W
     ssarily  T
      exceed
       ts ed
        resp
      urgent
     s pagin
    oward en
     concern
      re dil
       molog
```

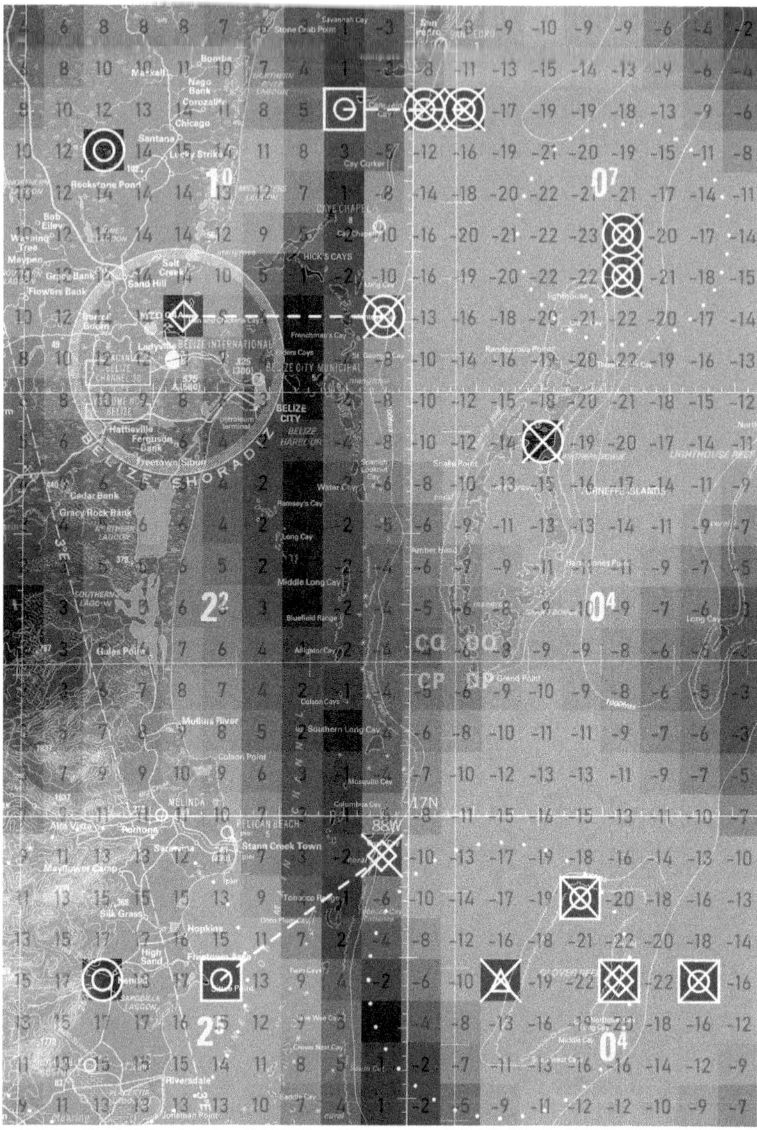

Meso-American Reef, TPC J-25C

e depletiTHER CONDI
ral resouETHER WITH
n increasNG MEASURE
ondensatiDUCED GAS
on dioxidE RESIDENT
r green COMMERCIAL
 in the AS SUCH GLO
due to DEMAND GREW
of foss STIMATED IN
 defore AS NOT ENOU
 increaAKE UP THE L
house g WHEN DEMAN
of the D BY DEMAND
what ha WAS PRIMARI
wn as a RTED BY CHI
 climate AMERICA AN
evidenceS RICH MARK
sing tem RICA AND T
requent E EAST CHIN
clones fED ITS POSI
 crop faTHE WORLD S
the melt LNG IMPORT
tain glacATURAL GAS
olar ice NCREASING
fferent p RAST NATU
e globe caNSUMPTION
ide is als FELL BY

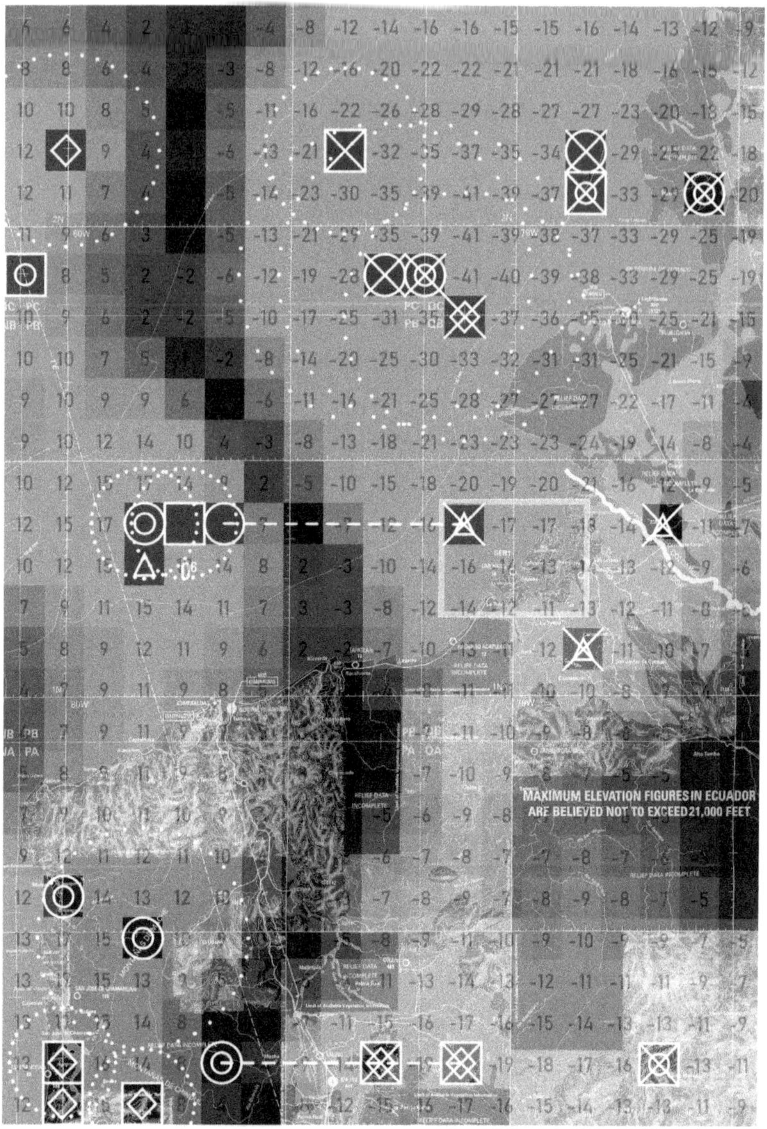

Tropical Deserts Zonobiome ZD-A2, TPC I-26D

mentsGING LIGHT AND
ined R FIVE TALL ST
mih LAMPS HAVE THE
ct mWN SOLAR SYSTEM
nery ELL GIVING LIG
tina ROUGH THE NIGH
ecoE VILLAGERS SAY
si c BEST THING OF
s hanBOUT THE ARRIV
er ma GREEN POWER I
nd des THEY NO LONG
amadi h ATTACKS FRO
ration BEARS FROM
ces depaOUNDING HIL
harge of DARK FLICK
tion dep INSTITUTE
es griff RIGHTS BU
macneill RB IS THE
y matthe AME OF THE
ola bayeTE FOR HUMA
rnbach t AND BUSINE
g hataicIMITED BY G
ehner oE INCORPORAT
ghligh NGLAND AND W
price EGISTERED OFF
opec DRESS C COMME
ket orROAD EASTBOUR

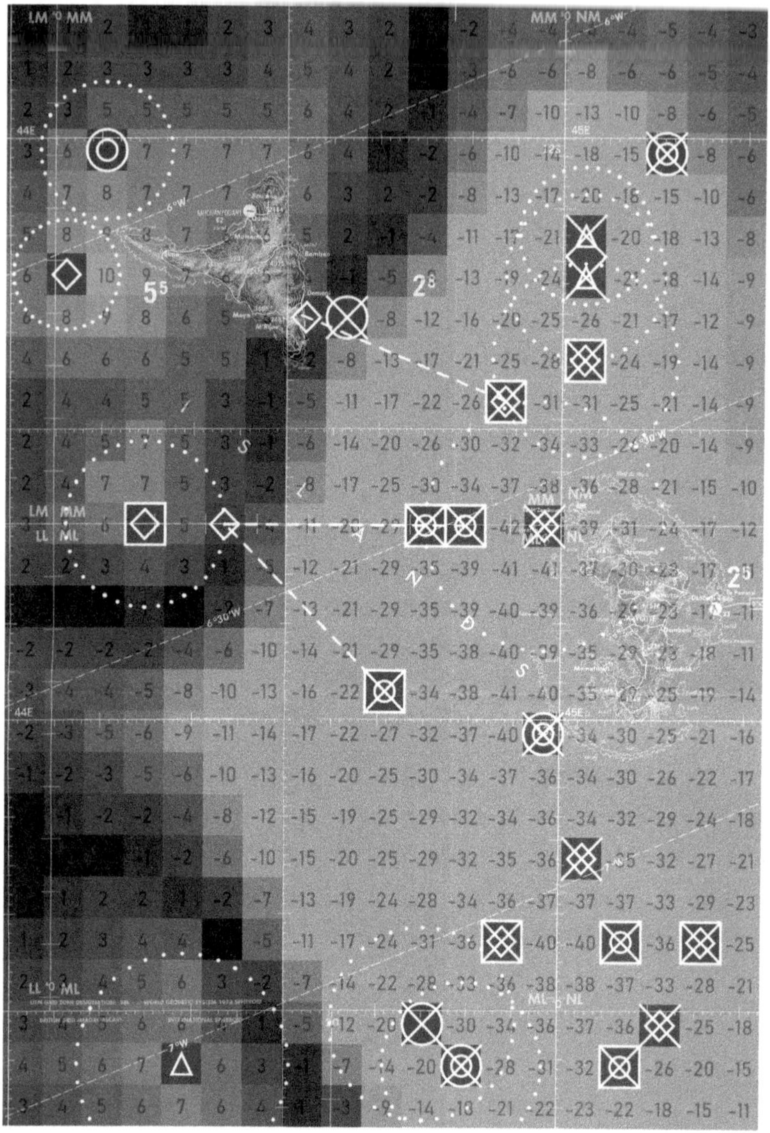

Back mangroves Tannes, Salt Meadows, Mangroves and Supralittoral Forests, TPC N-6D

ly demand OF PARTI
e its anal NOT HAV
covers coAL ADAPTA
s and miniING INSTR
ts at regiTHE RATE
country lSE DROPPED
al gradeR CENT TO P
s impacIN WHILE HA
supplyHE COUNTRIES
sions UT ANY SUCH I
pensabENT ARE IN TH
r thosESS OF DEVELO
nergy NE MOST OF TH
ssues PARTICULARLY
ERABLE TO CLIM
ATE IMPACTS AND MOR
E MUST BE DONE TO S
UPPORT THEM TO CLOS
E THE REMAINING GAP
FASTER FIGURE ES T
CONCEPTUAL LANDS
sionOF THE ADAPTAT
denn P REPORT SERI
ho proNECTING TEMPE
suggestCHANGE AND L
ments tF CLIMATE RI
e procesDAPTATION W

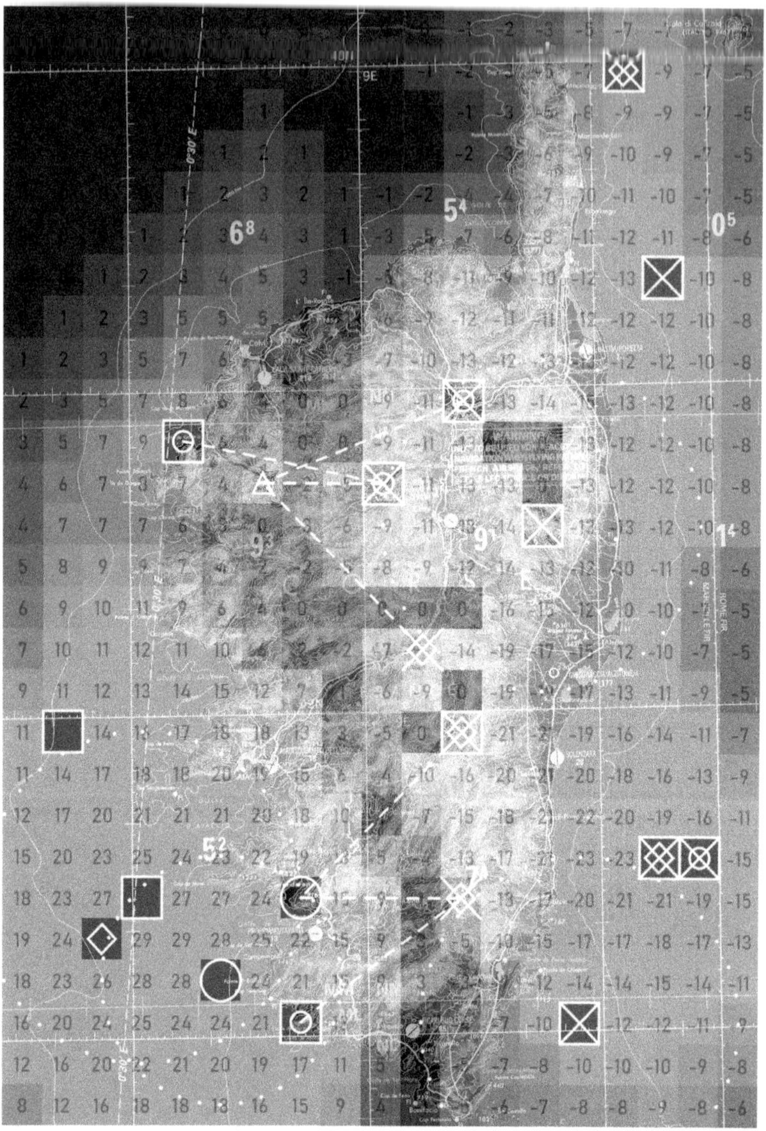

Nerium, Chaste Tree and Tamarisk Gallery Forests, TPC F-2D

THEIR PE
HE DISTIN
n   EN DEFOR
ver   D FOREST
the ION IS MAD
ntal pEARS AFTER
ate chTURBANCE OC
ing theY MEASURING
us of anMANENCE OF
erritor T DISTURBA
ea or o TIM IF TH
ities oT CANO Y IS
the d ED PERMANENT
f its fSHOWS NO SIG
boundar   REGROW
ion of sTHE THREE Y
panies orOW NG THE
does not C THE FOR
they areRBANCE PIX
or recommeNTO THE D
pcc in preON CLASS
o others of DISTURB
ar nature wSHOWS CL
not mentioF FOREST
dvertised tITHIN TH
of public RS FOLLO
print elecISTURBANC

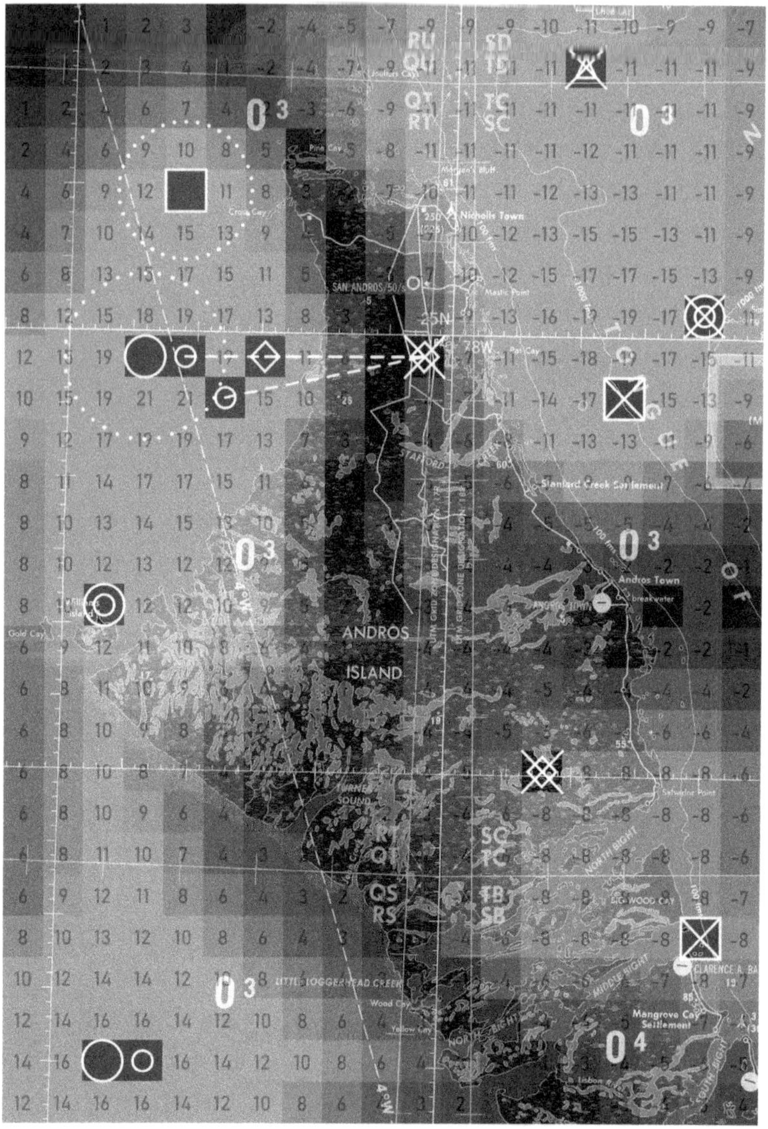

Caribbean Coastal Lowland Dry Forest, TPC H-25C

rep E BEEN THE BI
ntly CE OF SOME OF
a dry pEATEST INNOV
esult o EATS OF ENG
roductio AND CUTTIN
down anNTREPRENEUR
the foreHE WORLD W
lenging b ONE OF TH
hina pickiCREATORS
and in mid OF THE N
markets iLOGIES THA
as stablSIVELY CUT
s in the ON FOOTPRI
e to struOUR HOMES
tisfy the AND OF I
there areOTECT HARN
f heavy fTILISE THE
imposedND WATERS A
lowly cTHAT ARE ABU
nto the THESE ISLE
et as andCREATIVITY
acts are aISM IS NE
enmand forORE THAN
and belli TO AVER
stable the ORST IM
g of cites doATE BR
s by the foreORLD N

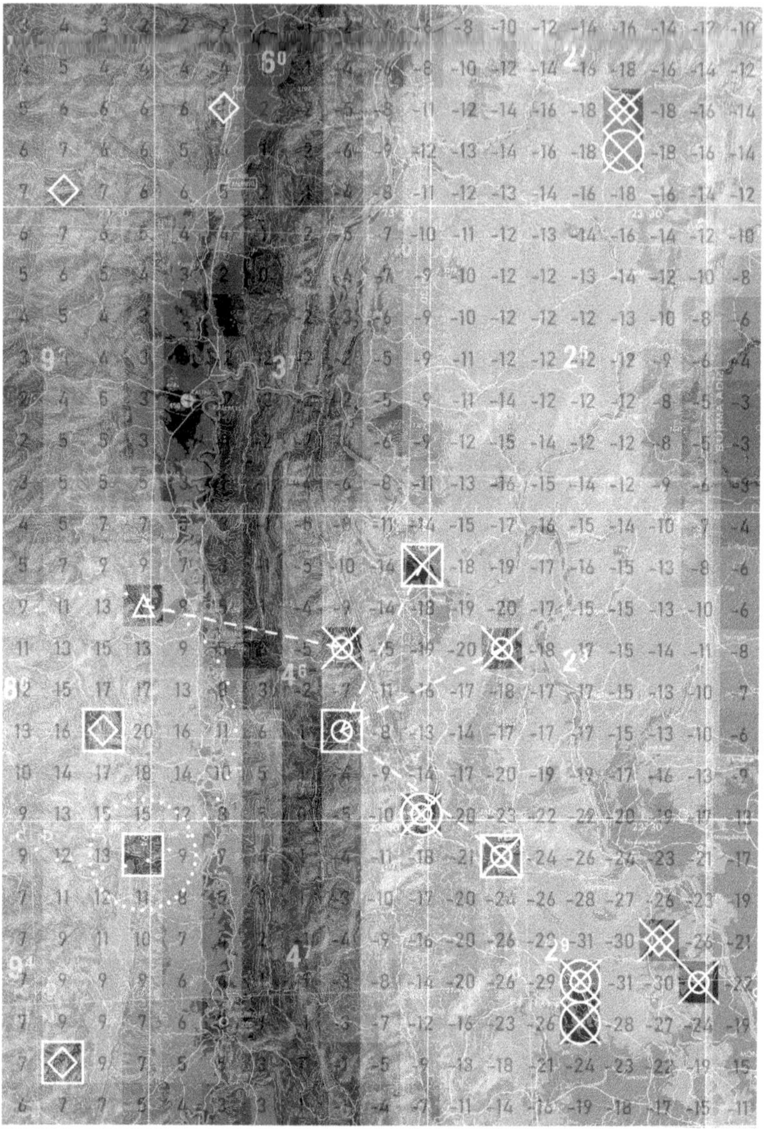

Chin Hills Warm Temperate Rainforest, TPC J-10C

s impliILL RESPOND
agricul WILL BE THE
d securIVE THEME FO
the natEAR AHEAD AN
ment inVEN WITH THE
ntries EGREE OF UNC
this r Y FACING THE
lly w TRY Q SAW THE
aga EST AMOUNT OF
nd oACTIVITY IN A D
es fo NEARLY MATCHI
cco is TOTAL VOLUME
rienciALS FOR Q LAS
ted dr MINING COMPA
e dryinNTINUE TO MA
f oasi ISITIONS AND
is incrMENTS KNOWIN
ng documHATEVER TUR
o a combTHEY FACE T
climate MATERIAL D
other huF THE ENERG
changesTION AND RI
the cliUPPLY DEFICI
projecSTARK STRIKI
acrossHE SURVEY RE
assa basIS THE LINK
criticaS PERCEIVED

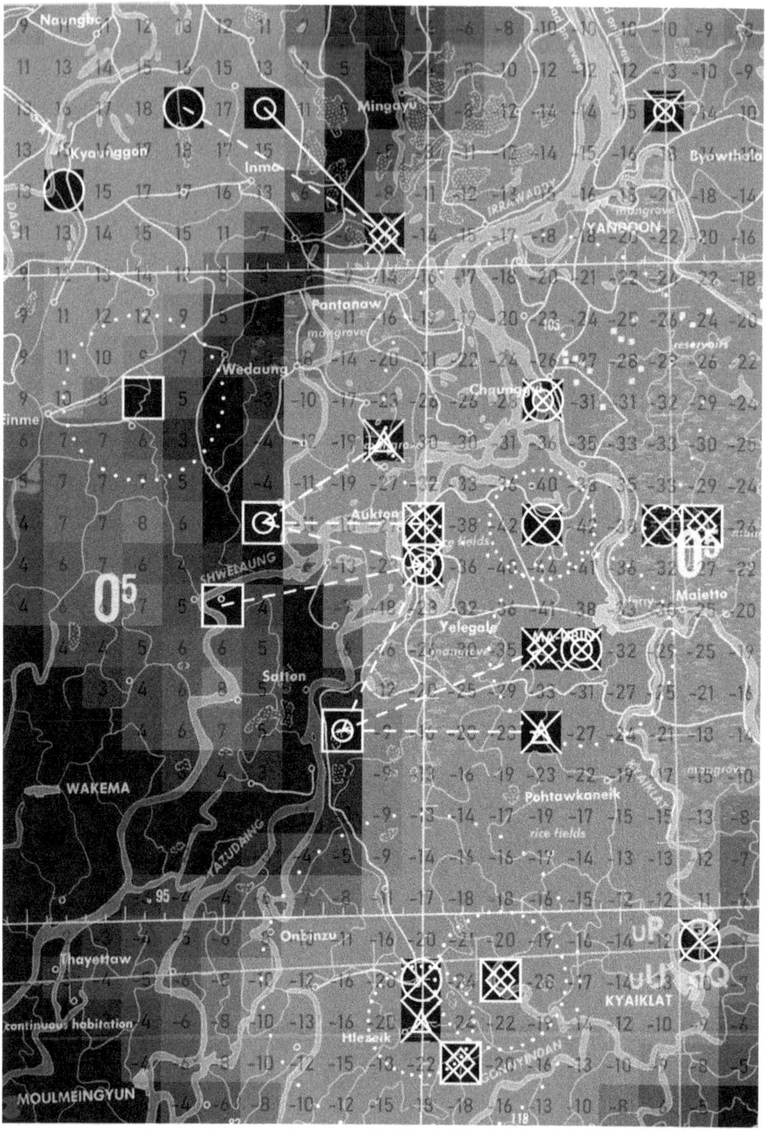

Central Ayeyarwady Palm Savanna, TPC J-10D

ave overtYEAR FOR V
s the topEVERY COUN
for the mGION THE S
ls sectorRN FROM MA
is littE TO MAINLY
ion thatMORE RECENT
s will ILLUSTRATING
inves SE IN AVERAG
s in mERATURES IN T
liftinUNTRY WWW SHO
uncertTRIPES INFO C
ahead VISUALS CREA
etals STATEMENT CLI
s stakHANGE HAS AN
to be ROBLEM THE I
ything iAT DEFINE C
or not rHANGE SHAPE
ell dam IT IS UNDE
john tND ACTED UP
o POLAR BEARS
LTING ICE AND ARR
YS OF SMOKE STACKS
DON T CONVEY THE U
RGENT HUMAN STORIES
AT THE HEART OF TH
ISSUE BASED ON IN
RNATIONAL SOCIAL

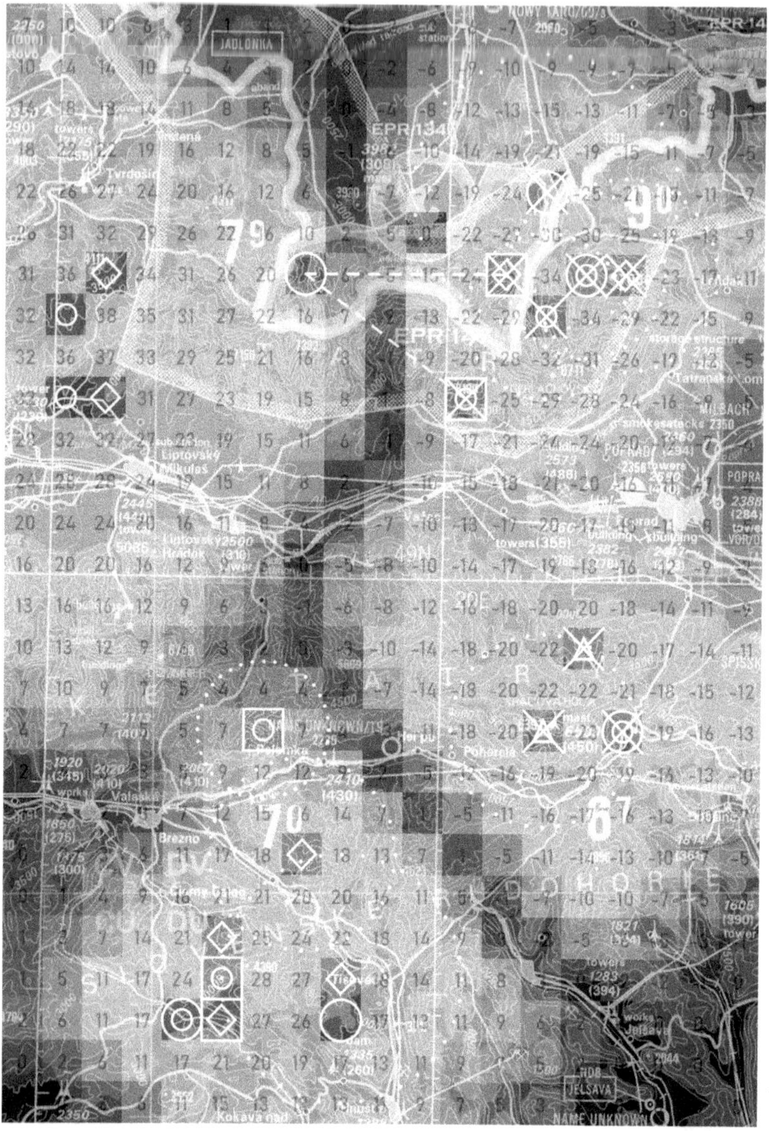

European Reedbeds, TPC E-3D

l energy TION IN TH
well as STATUS JUN
asis on IONS ANNEX
ciency iEARCH RESUL
 and theREST DEFORE
th in eleO EST DEGR
cle ev saND REGROWT
ople s reAMAZON STA
china herFOREST FOR
na was theCHANGE IN
or economON AND ITS
its strinON RAINFAL
 restricATURE CARBO
e end o ONS HEAT ST
 a postALTH AGRICUL
l deman E BIODIVERS
 the firsEFORESTATI
but deman DEGRADATI
n china slEDGE EFFE
dly from onFIRES SE
  global oilING ROA
 growth shriTION AN
om mb d in toNG FOR
kb d by nevert RE
s burgeoning pe
 emical demand
  rong consumpt

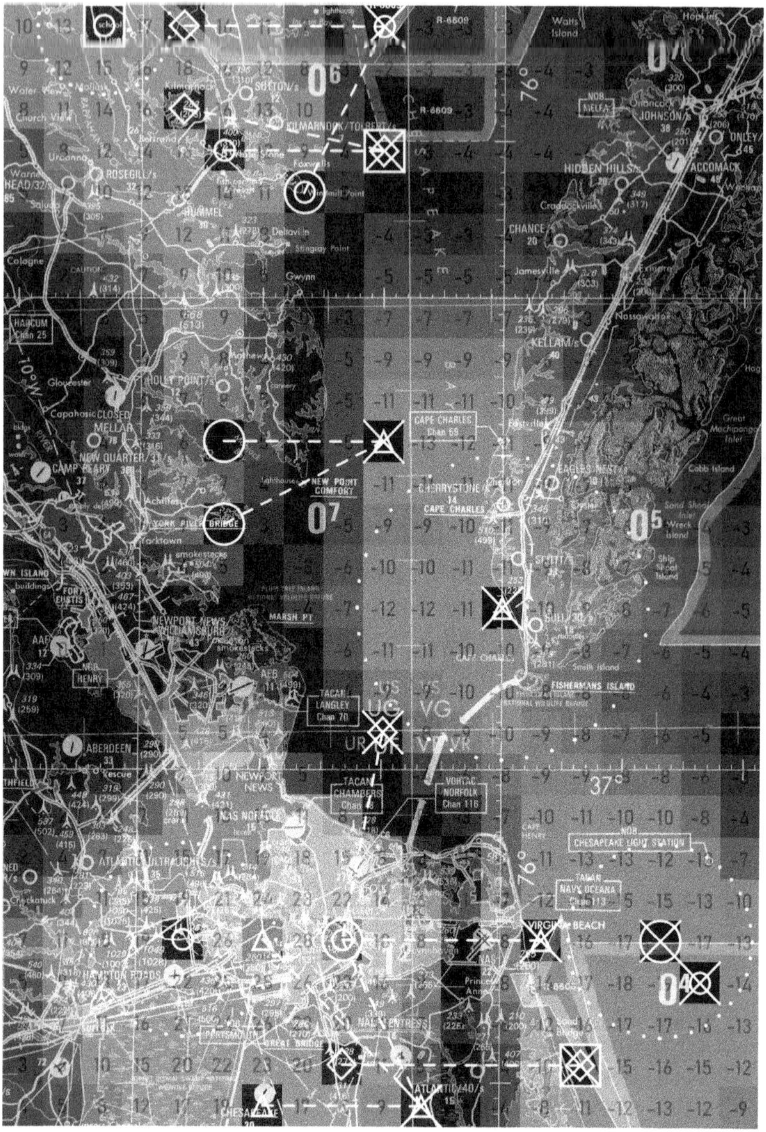

Chesapeake Bay, TPC G-21B

elimitatiHAVE EX
frontiersTURBULEN
ries the RS AS A R
specificHE PANDEMI
or produSIA S INV
ot imply KRAINE W
re endoERED THE
mende Y GLOBAL
efereSIS IN CO
of a HAVE SO F
re whicESS VOLATIL
entioned MORE TURMO
sed the LIE AHEAD
blicatiAL MARKET UP
lectr ICH PROVIDES
her fTEST ANALYSI
langAL DEMAND PR
d by TRADE AND
t extraNDS THAT C
is public AND SUPP
be reproduAN ALL T
out authori CONFIRM
rovided tha FORECA
te source i PROVIDE
y indicated Y ESTIM
al correspon FIRST
nd requests OUTLOOK

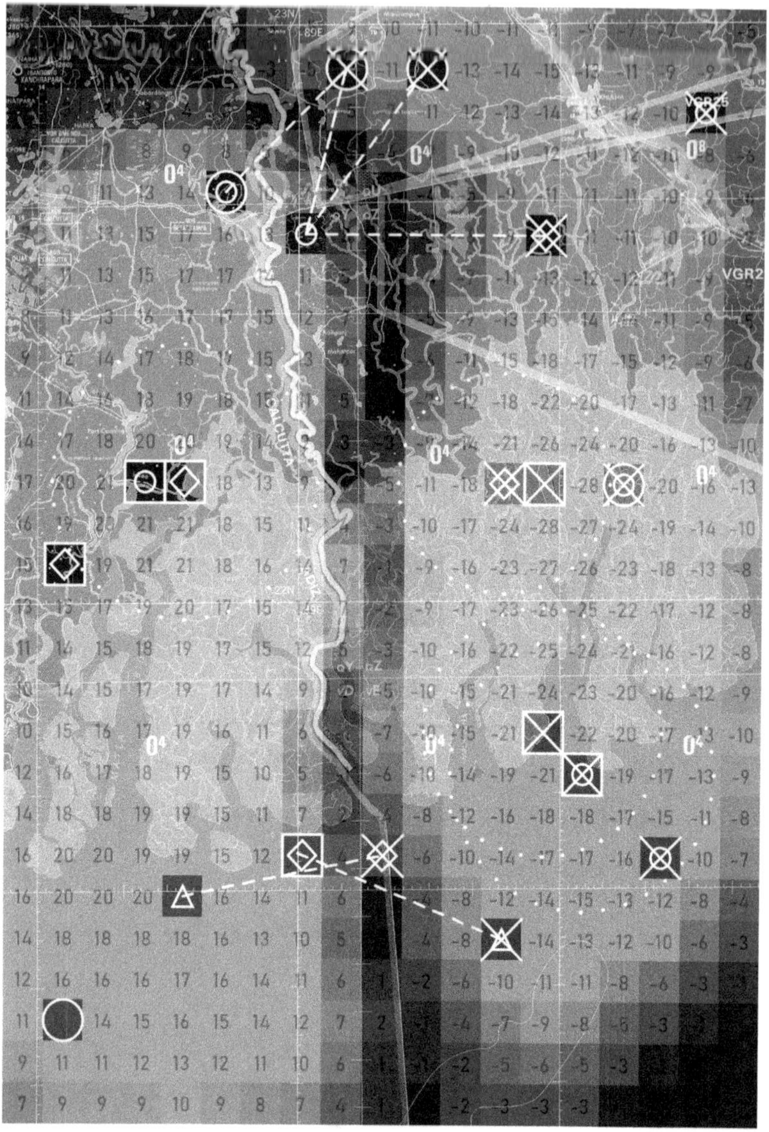

Sundarbans Mangrove, TPC J-9B

S BY THE FOREST A
THORITIES IS SAID
y jacqOING SMOOTHL
ra malayME OPERATOR
ichs madrPORTED THE
teven k r N REQUIRE
mina sahebOR THE BL
france robFORMS PRE
hez rodr gNT ON TRA
o diana rT NEWS SUG
hungary CER WHICH
china noIMBER TRAC
assaa algSUPPLIES T
ibuting aN THE NKOK
r s alegr ZONE CONT
hondurasFACE PROBL
ur usa biTHE AUTHOR
r friedl APRIL THE
rnelis bBLISHED A S
herlands ON THE SUS
iss switzF THE TRAC
ritania f BY MARC
k dentene IU AND JE
rlands sirS CHEVALI
norway erAINFORES
er switzer WWW AT
ory garner NOUNC

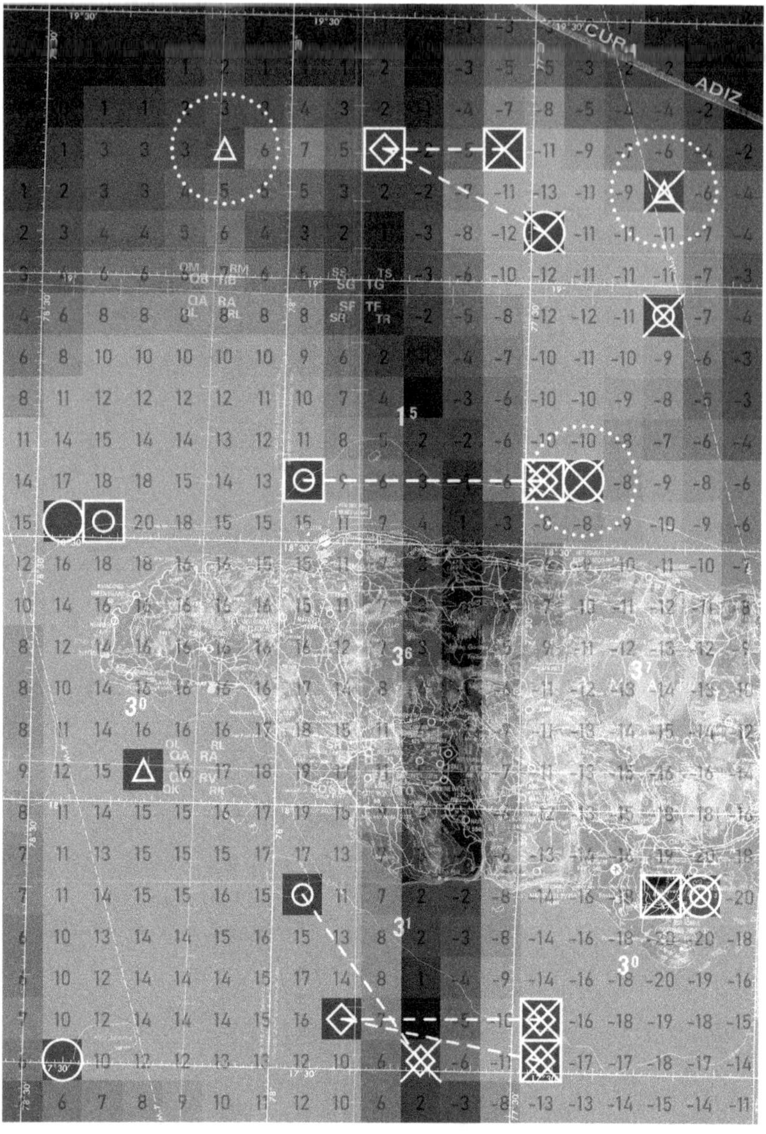

Caribbean Dry Limestone Forest, TPC J-26C

CREAT
d by t E WE N
ased on OUR WEI
t pipelinOUR INNOV
ects underNEERS AND
s light tiTO TRANSF
to growth ONOMY AND
ons we seeE NEED TO
net additi ELECTRIC
uction cap RT AND H
ught onlineEMS INTO
ite easing TURY WE
d on averageATICALL
o just kb d ENEWABL
capacity bu ENSURE
till moves TIES RE
with projecRDS BY C
nd growth oUSANDS O
forecast peT SECURE
producing ORKERS I
s outside tUSTRIES
alliance noCLIMATE
ominate medOMES FAC
capacity eOFFICES
plans wit WARM IN
supply booD COOL IN
the united WHICH W

Caribbean and Mesoamerican Seasonal Dry Forest, TPC J-25B

nts of the eSOUKIAF
rilemma securMAGES
fordability   ANGE
tainabilityNDS AND
 has long S OUTLINE
ffects on ESENT SER
  energy sCATIONS F
 heighteneTURE AND
n energy sITY AND F
ncreases URAL ENVI
 domesticN ALL SIX
ced renew EXAMINED
other noREPORT ESPE
els help EN CONSIDE
lerate thST A BACKG
ransitionOTHER PRES
ture of e EXAMPLE M
nd change ALREADY E
 importanNG UNPRECE
il fuels OUGHTS AND
replaced NG AND LOS
ng share   ECOSYSTE
le energy EASINGLY
creasing eMENTED DU
ation the tINATION
n to a low CHANGE A
orld requirMAN INDU

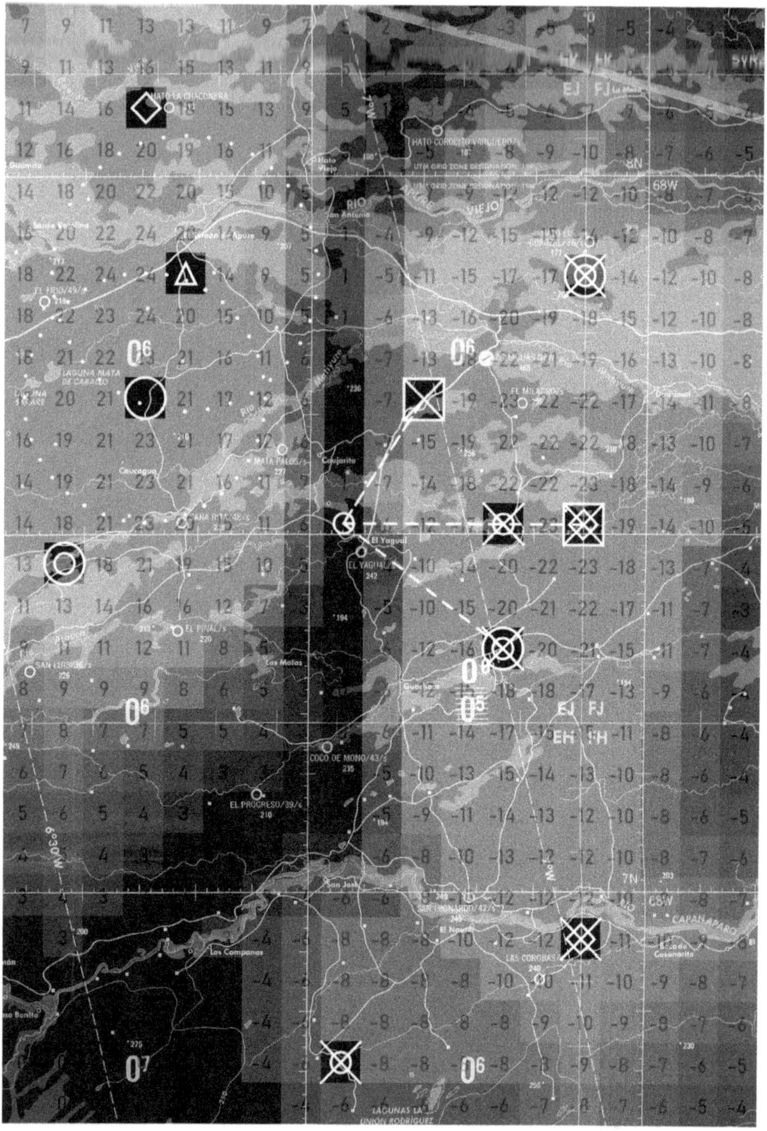

Llanos Seasonal Dry Forest, TPC K-27C

angmemba aAL IMPERA
g world ecT NOW CAN
joerg spitPUTED COU
researchE ALSO OBL
sd email PROTECT T
opec orgRIGHTS AFF
demand dCLIMATE CH
saad oilR INTERNAT
alyst ps IONAL AND
ad at op LAW FAILUR
d oil supTRIES TO P
akbar deNGEROUS CL
supply anNGE BY RED
email adeIR GREENHO
pec org MISSIONS W
kets and THE EFFECT
peration F THE HUMA
ba chief LEGAL FRAM
roducts CLIMATE CH
email CTIMS WILL N
pec o IN THEIR MILL
ets THE WORST EFFE
sen OF CLIMATE CHAN
cia ND ASSOCIATED H
i RIGHTS VIOLATI
HOWEVER CAN BE
NTED THE SCIEN

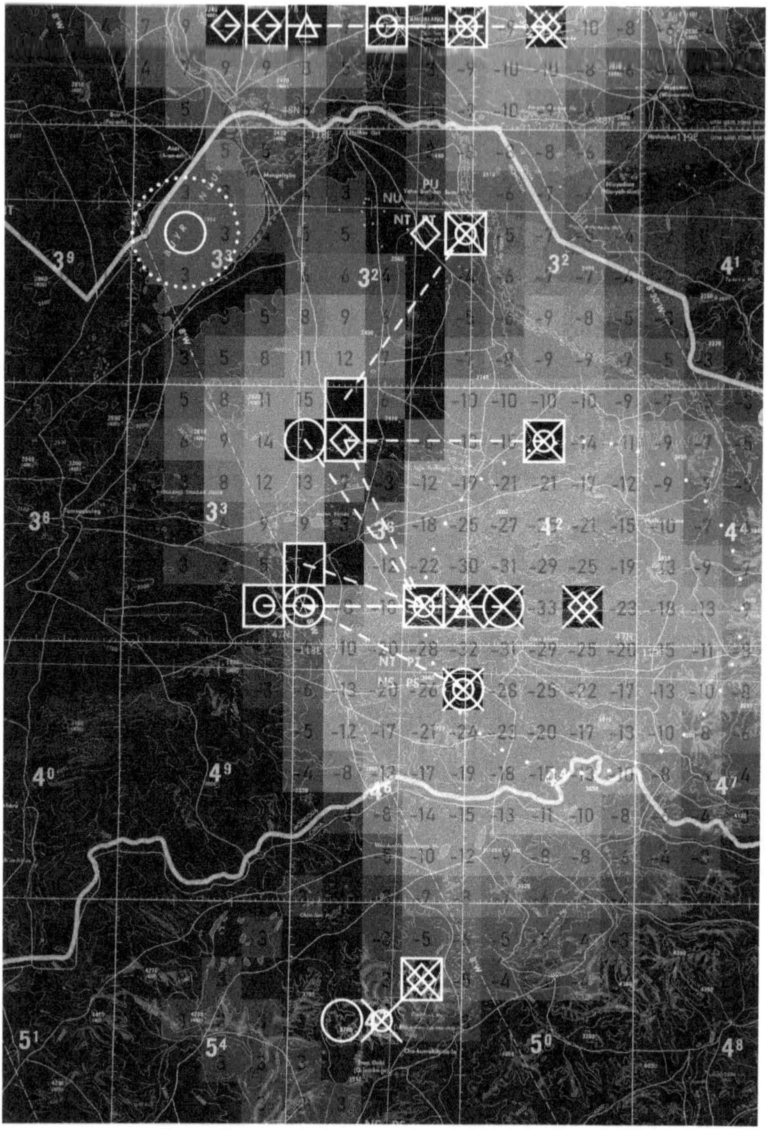

Sand Pioneer Community Ecosystem, TPC F-9A

in the mTITUTE IN
north afSENTED A
content TO JRC
it kha NUMBER
fp via SALS T
s the cliLIAN GO
related RECENTL
projecti INTO TH
d above RATIFICA
rious imSS ARE PR
or agri SECTION
ood sEPORT IN P
r the THE BILLS
ronmeDEMARCATION
co DIGENOUS LAND
i LAND GRABBING
ENVIRONMENTAL LI
ING ON LEGAL MI
IN PROTECTED A
ND THE PROPOS
ccoION OF THE
erieN LEGAL AM
ented BETWEEN
the E FIRST
ss of oEROUS N
ms is ART eL
eing H u

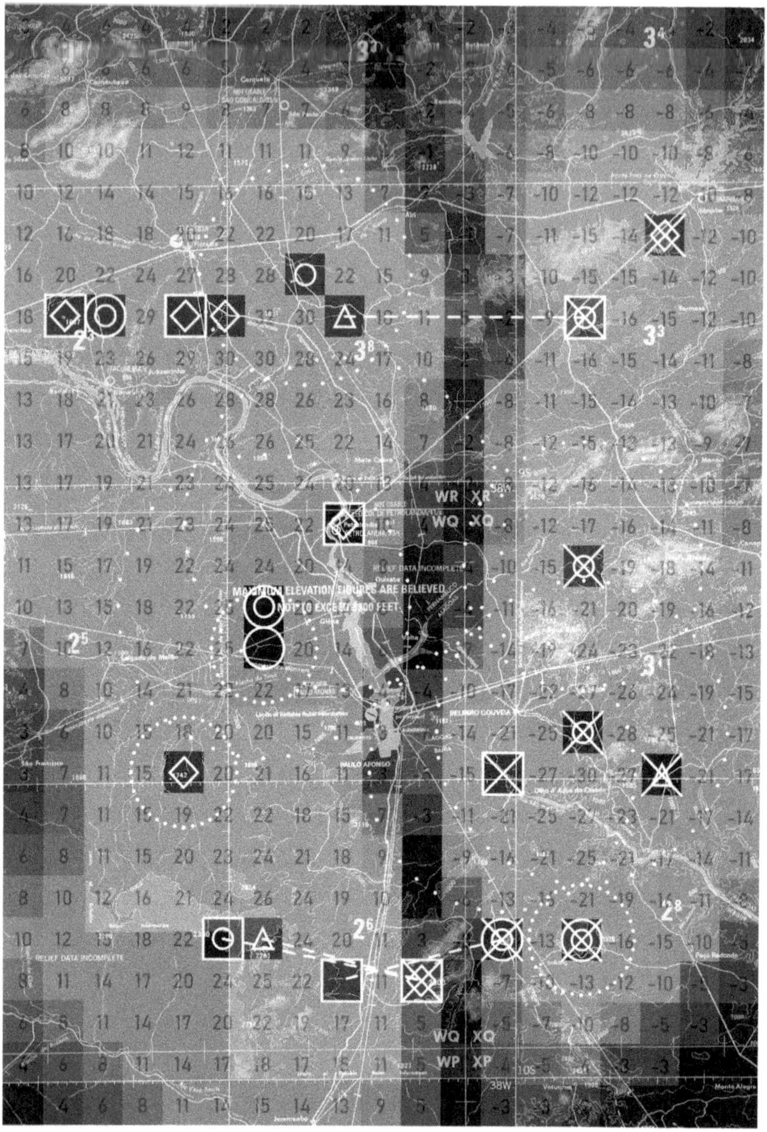

Caatinga Seasonal Dry Forest, TPC N-28B

ponsible ULAR BUT
it also e NDIVIDUAL
jor trends MEMBERS
vidual min FIVE ARE
discusses SSEMBLY
cy implicatONGLY SU
report willTURE IN
lowed by a fESSES M
ing analysis USING
ill feature ER CARB
ensive demaND MATER
upply projeACKED A
or key mateECIFIC P
d a number SUPPORT
dives on keCLUDING
it also m FICIENCY
lable an oND STANDA
l the critNDED PROC
rals data AWARDING
which allo CONTRACT
o explore ES PREFER
vely the la CARBON
projectionND PRODU
of content N PRODUC
tion execut AND SE
ary key markXTENDE
lopments devSPONS

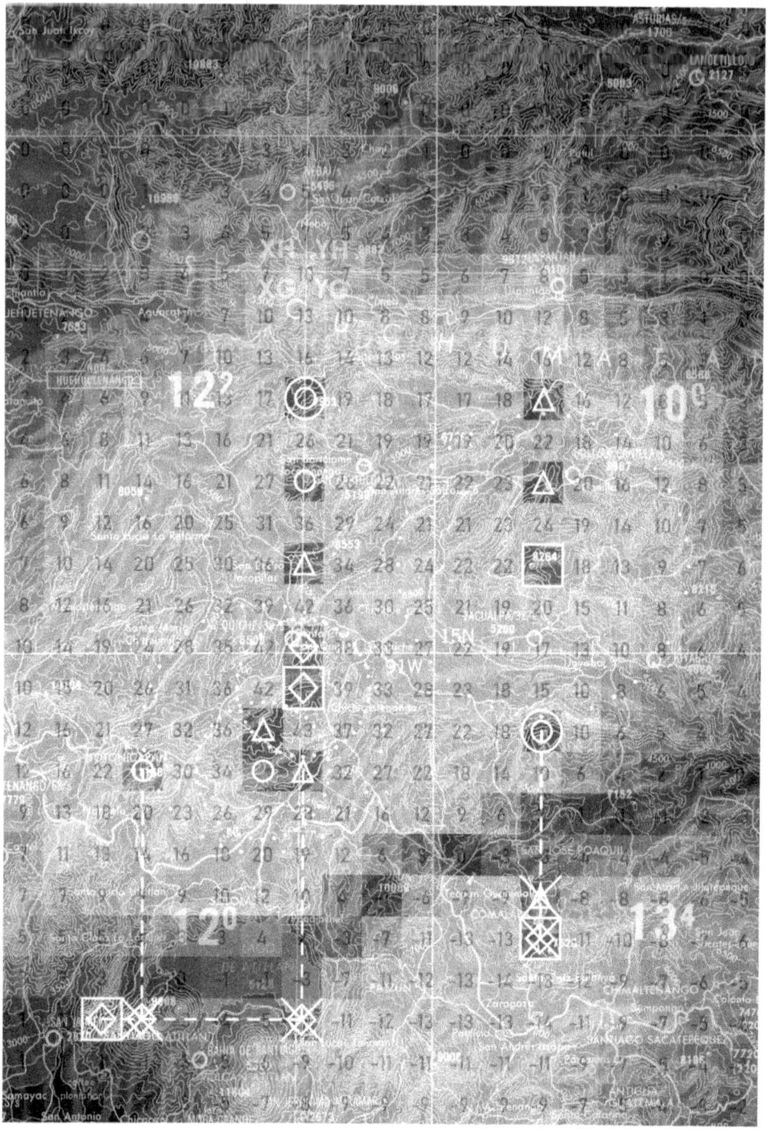

Mesoamerican Submontane Humid Forest, TPC K-25A

```
       i
       to
     razil
    rowth i  i
   d up to and
  ected to grow
 n russia s econo
c growth in is rev
sed up to and a fur
ther recovery is an
ticipated for with
a growth forecast o
f world oil demand
world oil demand is
 expected to grow b
y mb d in following
 an upward revision
 of about mb d from
 last month s asse
sment mainly du  HT
 higher dem RC EC E
 in chinaTMF EXPLOR
 americaTTPS EARTHE
d up PARTNERS APPSP
couCOM SCIENCE GLOB
thaFOREST HTTP WWW
rma INPE BR OBT ASS
```

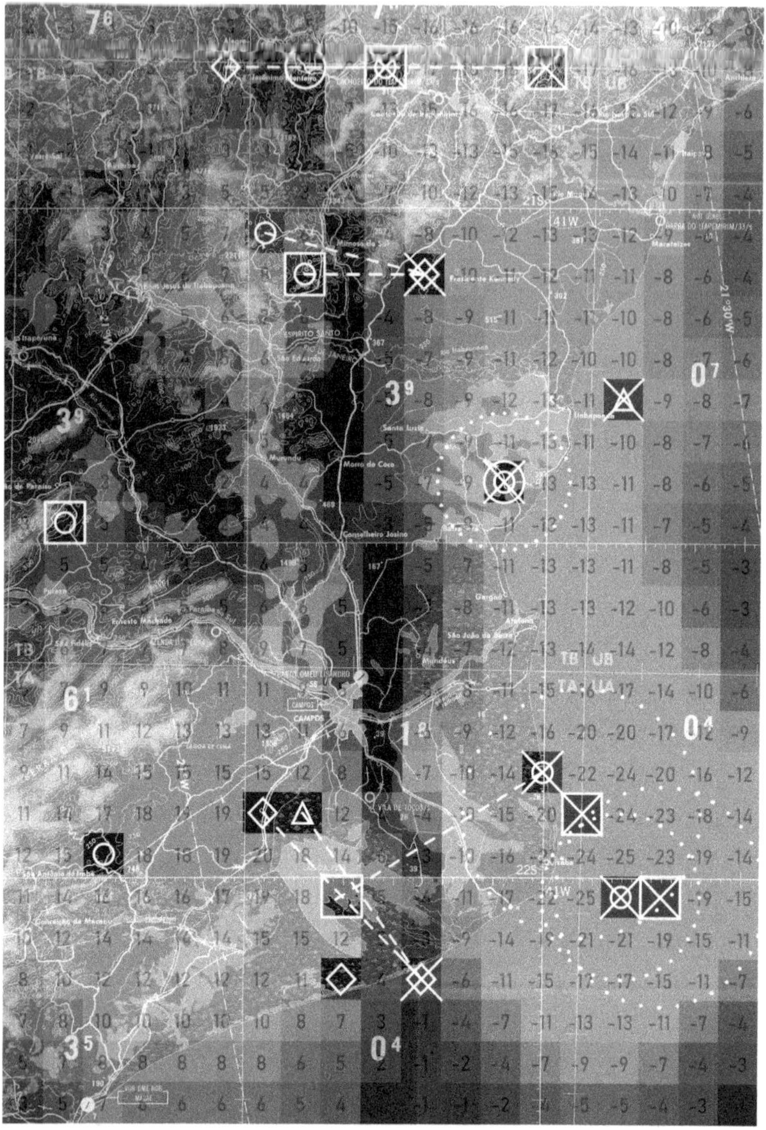

Brazilian Atlantic Montane Humid Forest, TPC P-28C

DO NOT IMPLY THE EX
PRESSION OF ANY OPI
NION arATSOEVER ON
THE k is OF THE INT
R ryoneMENTAL PANE
avigate ATE CHANGE
tain futuNG THE LEG
lerate t OF ANY CO
n to g RITORY CIT
r as aA OR OF ITS
eed ack TIES OR CON
ook nd THE DELIMIT
be improvITS FRONTI
e most welNDARIES T
cer dale c OF SPECI
omist welcIES OR PR
e edition S NOT IMP
nergy outHEY ARE EN
beliefs t RECOMMEND
outlook c IN PREFE
d to identTHERS OF
ts of the NATURE WH
ansition tT MENTION
ommon acroRTISED TH
in scenari PUBLICAT
trends helNT ELECTR
ore beliefNY OTHER

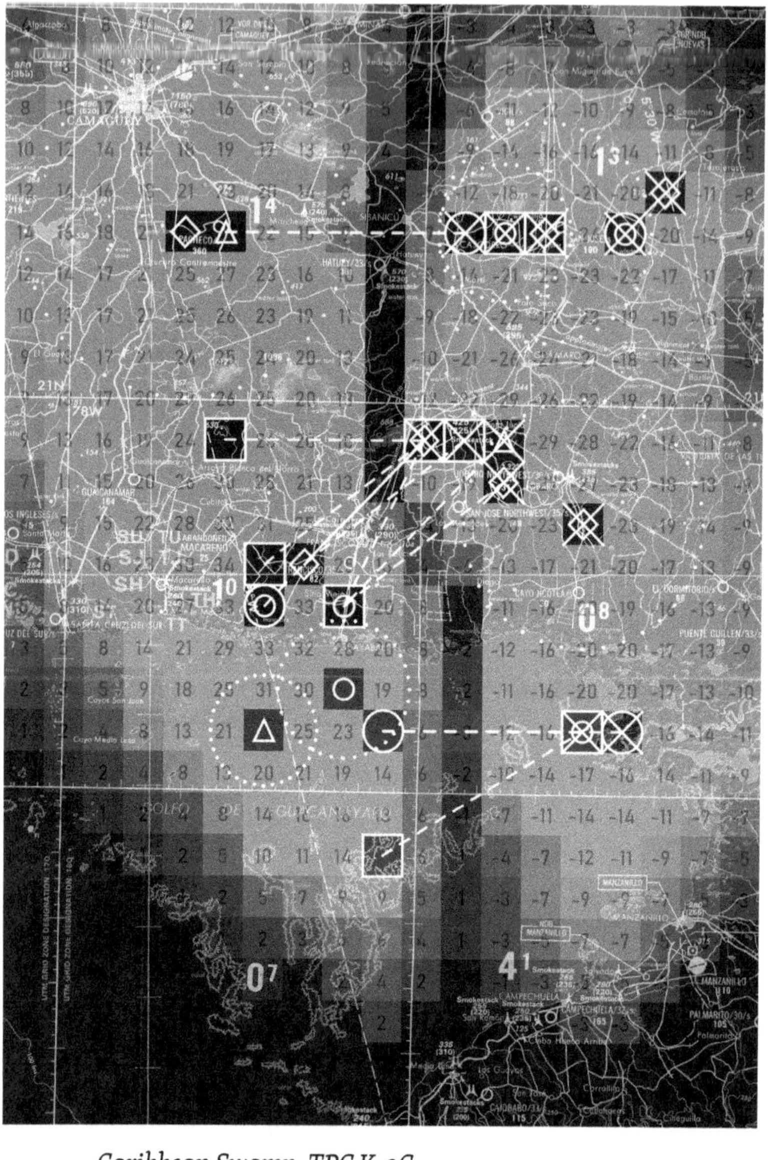

Caribbean Swamp, TPC K-9C

ed to harmoTO ACKNO
surement moEBT TO T
 reporting EORISTS
fication mKING IS T
 and incen FOR THE
e investm TION OF M
red to ef BARAD KAT
reduce natF ELIZABE
related grELLI SUSA
gas emissi DONNA HA
missions IDA NEIMA
fit from sT THE POS
icy moment ESTAB LI
issions gasD NOT HA
nued to benNE FINAL
m a wide ra OF ALL
olicy initiHANK MY
n the europY WAGNER
  launched A SOURCE
  ncing mecED INSPI
   hydrogen VERING
    politicaIMMEASU
     was reaOR LIFE
      y rog EVER M
         e   NCE
              AL

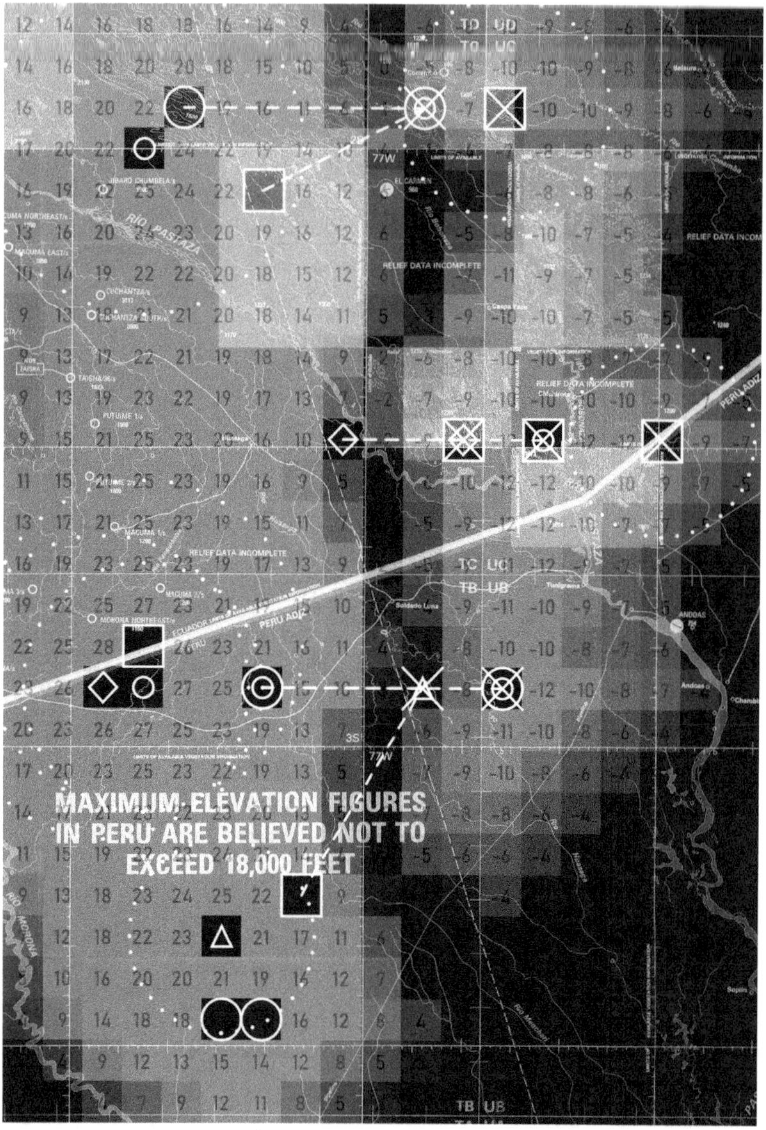

Western Amazon Floodplain Forest, TPC M-25A

remains ORMED BY
is expect TERATURE
nd by in bPIC AN EQ
conomic grANSITION
s revised GNIZE T
is expect S CIRCU
w by in ruFFER WIDE
onomic groNG ON THE
revised uAL AND IN
a furtherAL CAPACIT
is antici AS THEIR
with a gr OCIOECONO
ast of worDENCE ON
mand worldL PRODUCT
nd is expeON THESE
row by mb ONE MIG
owing an uIGHER INC
ision of IES AND
from las DEPENDE
assessmenL FUEL
ue to hig LEAD
seen in cN WH
oecd ameriTY
evised up s
to account fo
tter than ex
erformance

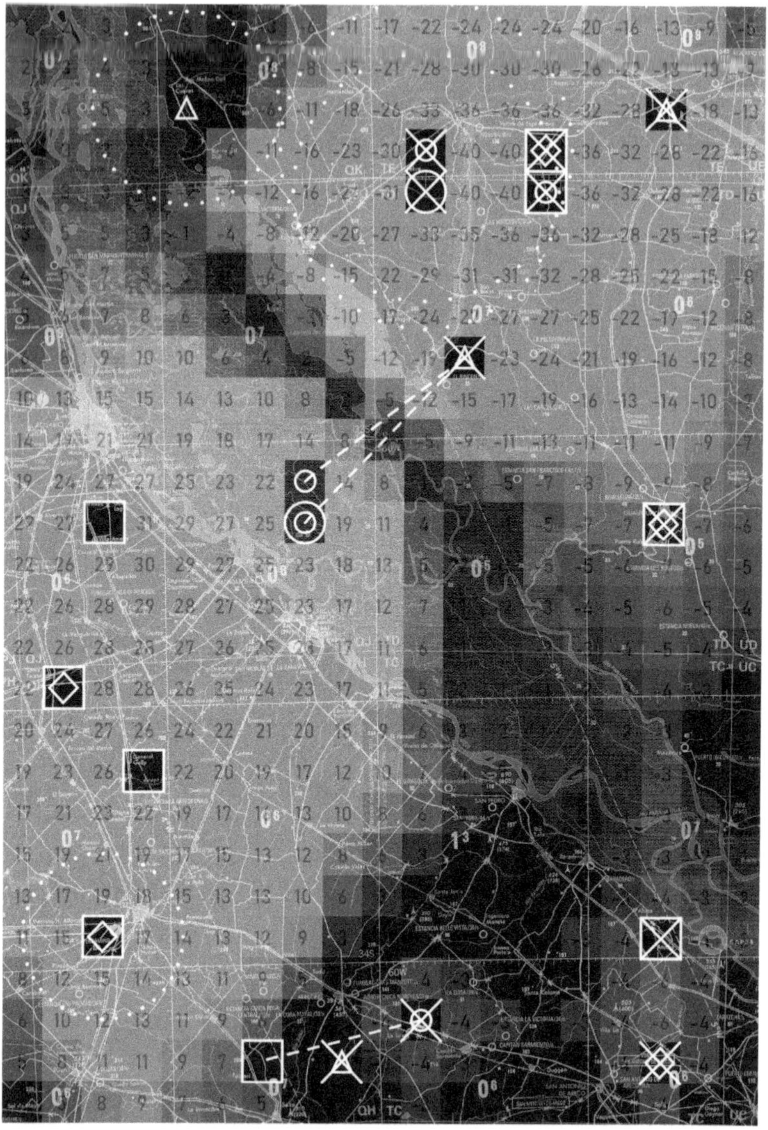

Parana Floodplain Forest, TPC R-24A

found IONS OF THE H
semblRIGHTS FOR MIL
highe OF PEOPLE PAR
nsidLARLY THE POORE
peedEOPLE IN THE WO
al prHE CLIMATE CRI
ce th LL CAUSE INCR
world LOODS DROUGH
costs ofS AND SEA L
 compared LEADING T
atives freON INCREA
ers and thoRANGE AN
fly further DISEASE
pay more staNG CURR
titive and pROBLEMS
the economy ER SCAR
the populatioIVE DI
aking the neceECTIN
 changes take aFOO
nt of differen ENC
vel needs e g TO L
e with fam   E AROU
way promo LD MOREOV
entivise LITICAL IN
s scrap Y ARISING F
to make MASS DISPLA
more e gF POP LATIO

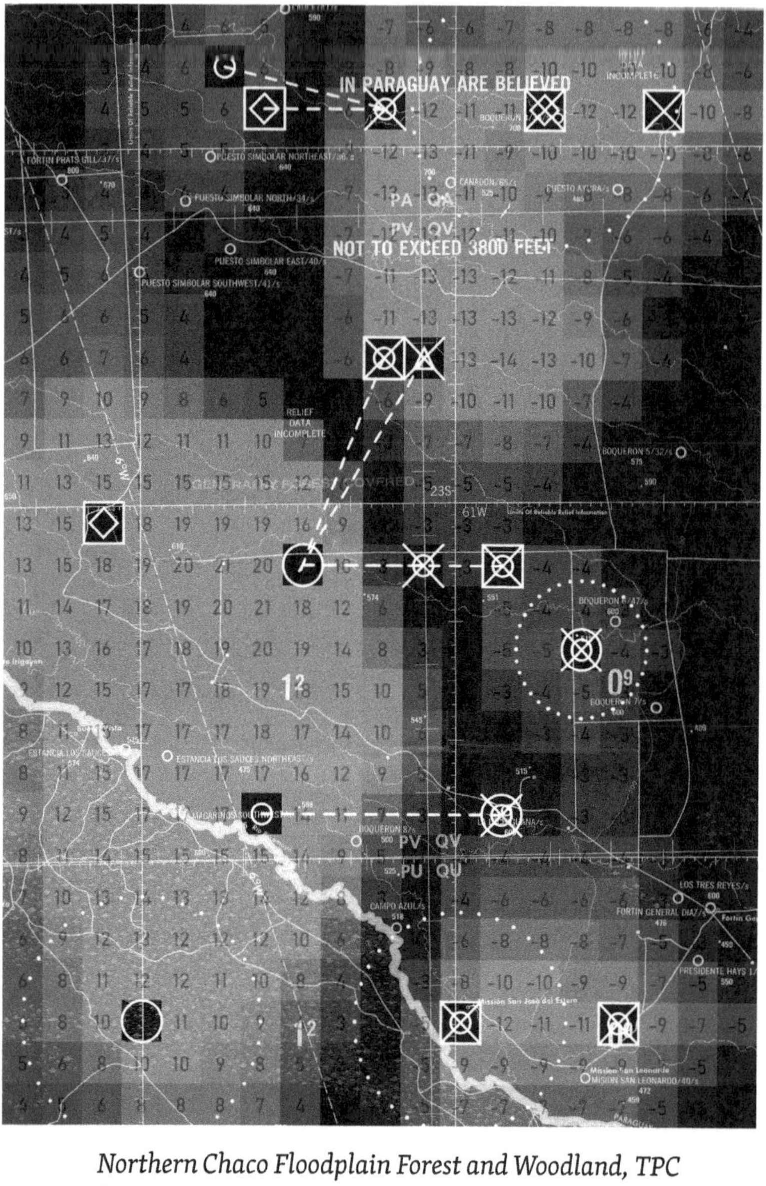

Northern Chaco Floodplain Forest and Woodland, TPC P-27D

a algeAST YEAR HA
ting auOMINATED BY
alegrRRIBLE CONSE
onduras OF THE RUSS
usa biINE WAR AND
friedl UL TOLL ON L
elis blD COMMUNITIE
rlands HOUGHTS AND
s swit RE WITH ALL
tania frFECTED FR
dentener Y PERSP
ands siriISRUPT
orway ericN ENER
switzerlaND THE
y garner usBAL ENE
guivarch fS SEEM L
rjolijn haasE A MA
e netherlandSTING
t hansen ger ENER
thias hauseBAL ENE
rland ed ha AND DI
tim hermaN RECENT
therlands BEEN FOC
pp usa no E IMPORTA
nce ringuCARBONIZIN
jared lewRGY SYSTEM
ia new zeTRANSITION

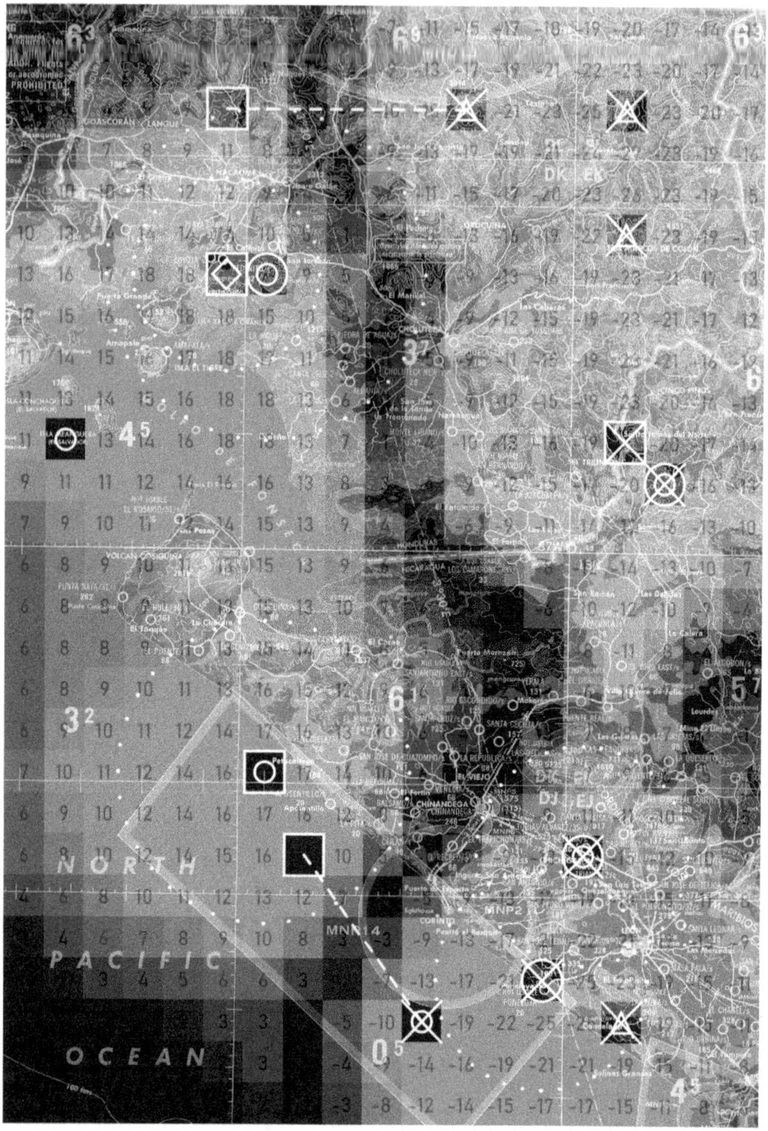

Eastern Pacific Mangrove, TPC K-25B

weath TS TEMPER T
recent NG IN GLOBA
ight morPPLIES THE
han everIMPROVES OV
ance of ORECAST PER
shift towA S INVASI
zero futuINE SPARK
ents of th IN OIL P
ar have h BROUGHT S
the comp SUPPLY C
interconn THE FORE
of the glCCELERATE
gy system OF CLEA
eed to addrCHNOLOGI
three dimensME TIME
the energy ESTMEN
a i hope thisECTED
s energy outlHEIR H
useful to eS SINCE
trying to TIONS AS
this unce OIL PROD
ure and acTAIN THEI
the transO BUILD UP
obal neY EVEN AS D
M ys a OWTH SLOWS
R he oING SPARE C
C t cCUSHION OF

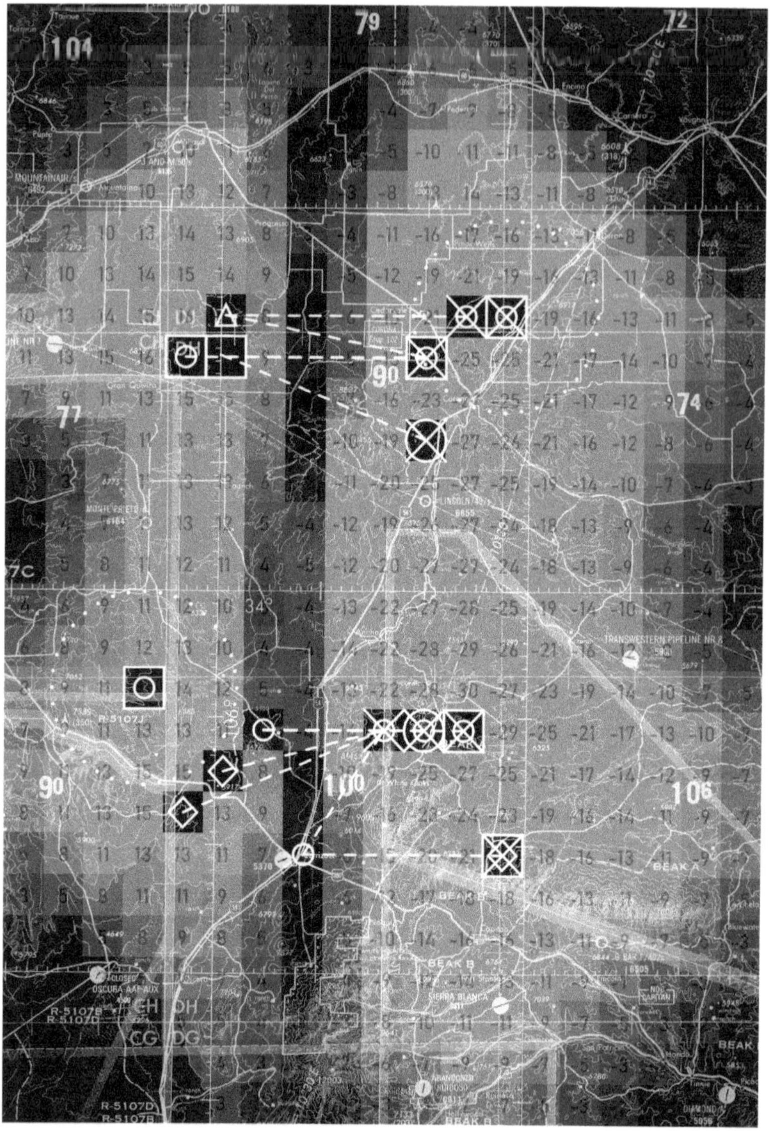

Southern Rocky Mountain Two-needle Pinyon - One-seed Juniper Woodland, TPC G-19C

ccelera SEEN
bal ada OECD
ion currREVISED
action TO ACCOU
inadequETTER THAN
the tem PERFORMANC
d adapta US IN Q SI
of the pOECD EUROPE
ent whilSED UP SLIG
erage teQ IN THE NO
are alrDEMAND WAS A
ng c abISED UPWARD
strial UNT FOR BULL
nt plan DEMAND SEEN
in the NA IN Q AND
eterminT IMPROVEMEN
ions ndTIN AMERICA
ng us oE SAME PERIO
ards c RLD OIL DEM
of the ORECAST TO
n if theA HEALTHY M
mperaturHING ABOUT
y slows OECD IS AN
of moreD TO EXPAND
ollecti WITH OECD
ange m CONTRIBUT
rts cARGEST IN

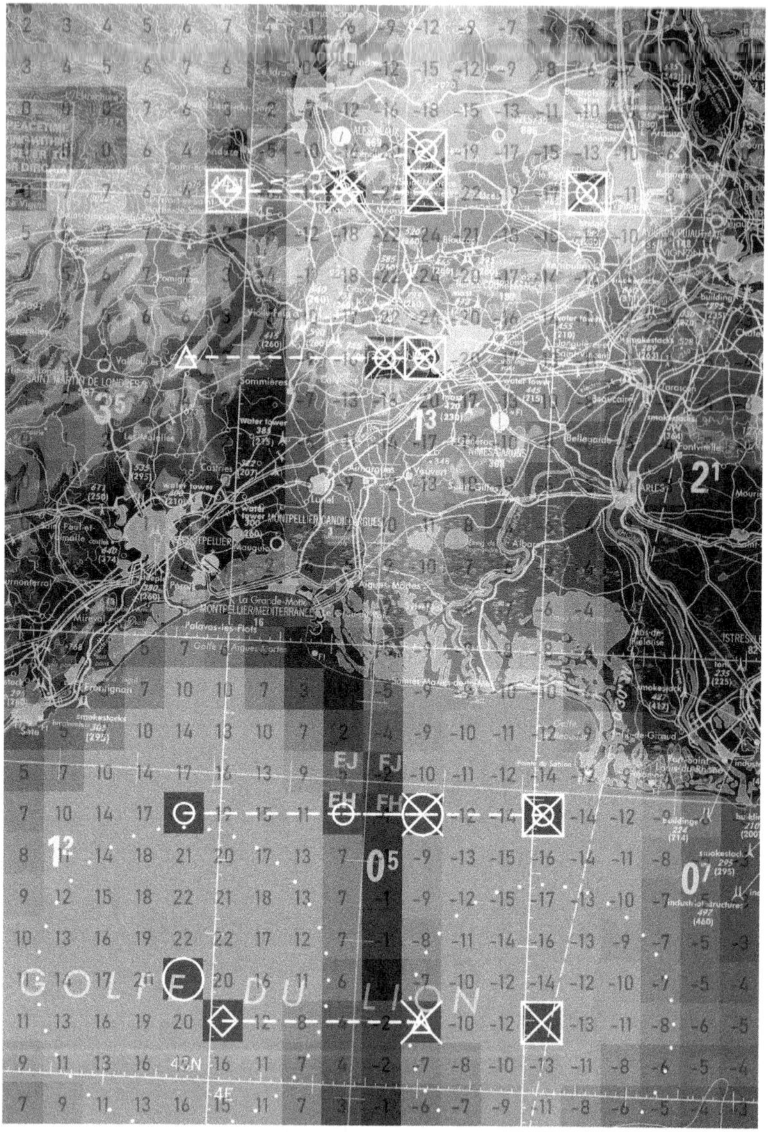

Mediterranean Wooded Dunes, TPC F-2D

trees aNE IN RU
egovic GAS DELI
yri EUROPE BCM
ud DUCTION GROW
the SHORT OF PRE
nding EXPECTATIONS
ick la A COMBINATIO
etitivROJECT DELAYS
opensEEDGAS SUPPLY
des ofS THE UNITED
rough ACCOUNTED FO
reneDDITIONAL LNG
oveY AND BECAME
traRLD S LARGE
tualXPORTER TH
anthr OF MARKE
it force S IN WAS
ary and gDRIVEN BY
consideraD SIDE THE
western PANSION OF
smailbegES AND IMP
s many ofCLEAR AVAI
ralities WEIGHED ON
elevant f AS DEMAND
ing a polE AND MATU
ime in thS IN ASIA
cene suchRICES LOWE

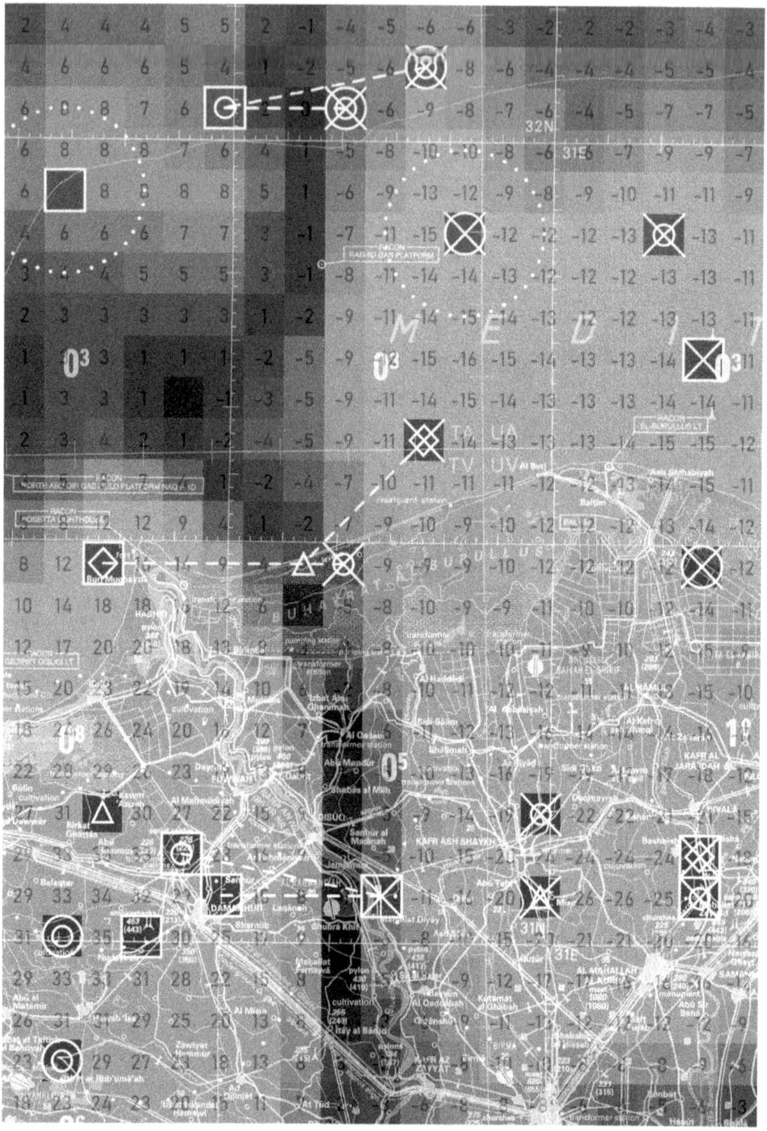

Sand Plain of the Lake Burullu, TPC H-5A

technoS THERE IS S
as solaNGLY LITTLE
teries T ON WHAT T
g unprecMIC OUTLOOK
wth in tMARKET PLAY
mineraL RESPOND OR
lectricILL BE THE P
ncreaseE THEME FOR
eding AR AHEAD AND
ene VEN WITH THE H
tems EGREE OF UNCER
en mor FACING THE I
h with Q SAW THE L
itions MOUNT OF M
solar p TY IN A DEC
ons contLY MATCHING
tter preAL VOLUME O
ds and w OR Q LAST
s set to NG COMPANI
upward mUE TO MAKE
two subdIONS AND I
this hasTS KNOWING
ignifica EVER TURBU
in dema Y FACE TOD
ical mineTERIAL DEM
to demandHE ENERGY
energy seN AND RISK

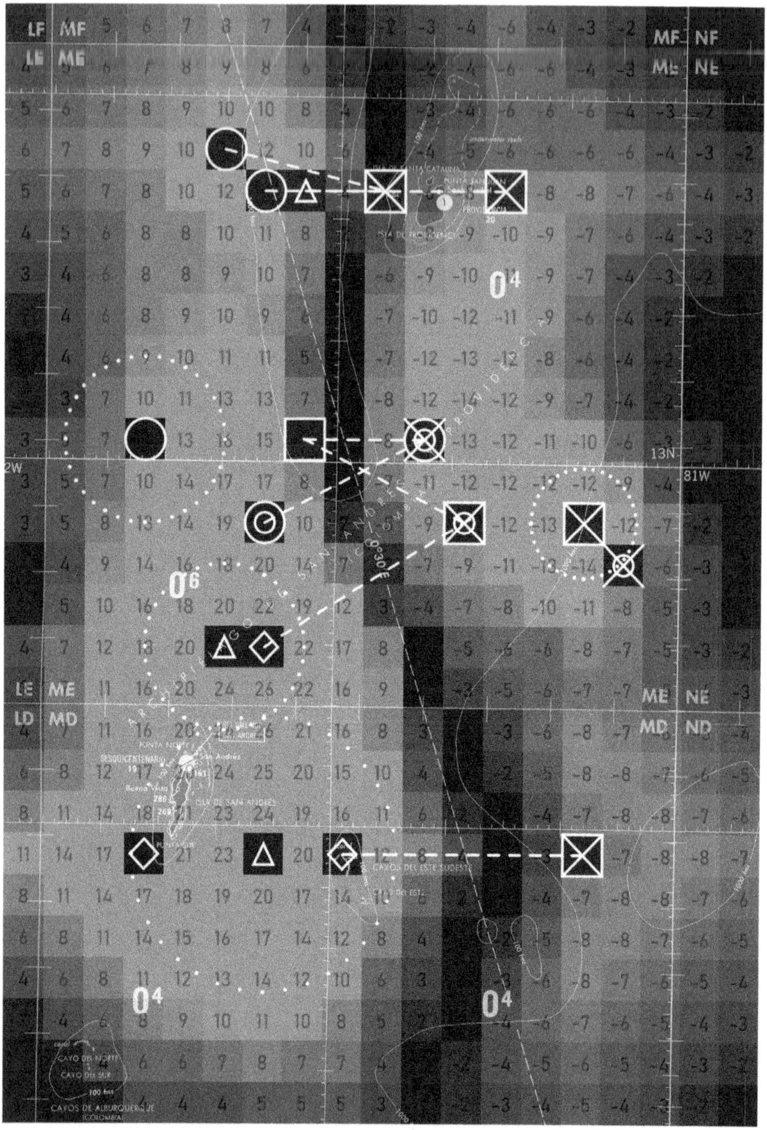

Oceanic Caribbean Sandy Beaches, TPC K-26A

tourist aELECTRIC
erted andION FROM
orator an NS SOURCE
le to par CLUDES NU
t anywher RENEWABLE
obstructi SOLAR WIN
 come he RO IS SET
a film a ALL GLOBA
nkholes GROWTH OV
een ravaT THREE YE
ad sea c ISSIONS SO
er the l CH WILL R
ears the ROLE OF F
are causeS IN PRODU
opogenic iRICITY GL
ons into thFORECAST
geophysics FOR ALM
rea where t THE WOR
extraction oCITY GE
als and the UP FROM
on of water NEXT T
ver jordan tOW EMIS
ate desert oTION IS
 has lowere AT TWI
a level leaAL GROWT
the creatioEEN AND
 ties unde TIAL CHA

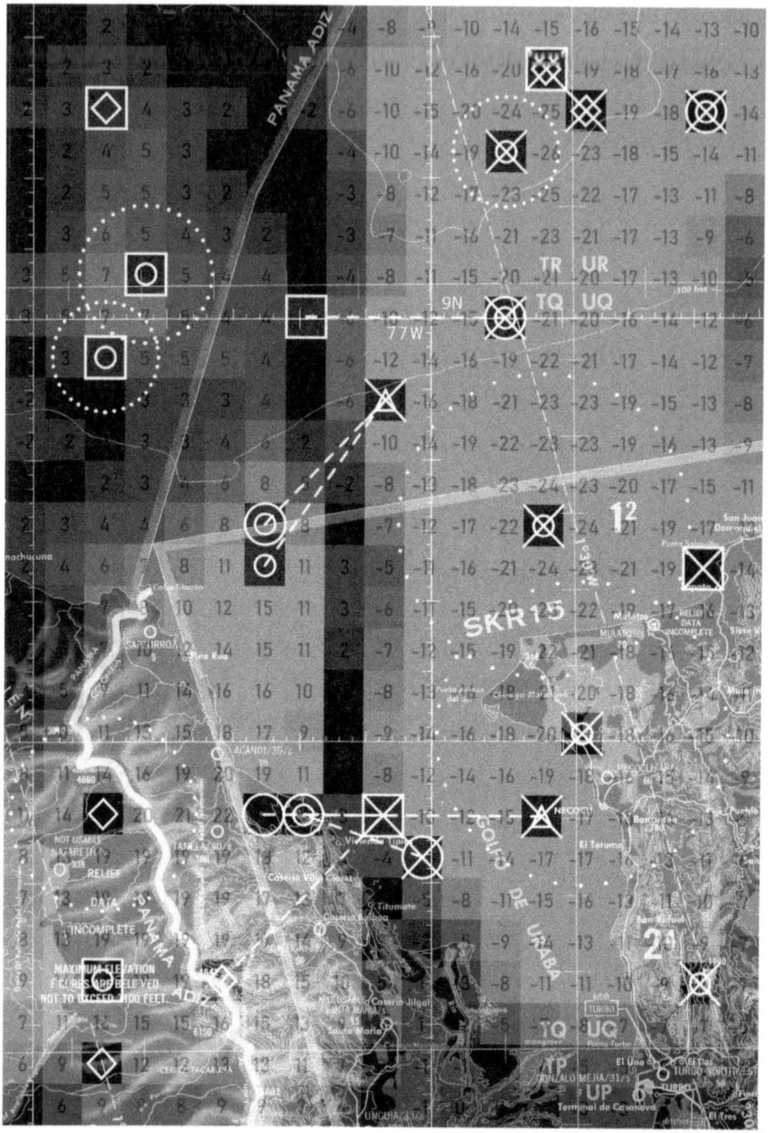

Continental Caribbean Cohesive Rocky Shorelines, TPC K-26D

rbo z NS WILHEM S
e poli RWE PAUL SI
and en LE UNIVERSIT
ewables IRA YABUMOT
er thanR THE INDIV
sil fue ND ORGANISA
long t AT CONTRIBU
imized HIS REPORT
cenario RESPONSIBLE
F anal OPINION OR
Jeport NT IT CONTA
I ar daERROR OR OM
est that THE SOLE
arming toILITY OF T
coal oil R QUESTION
oduction MENTS PLEA
line rapiT CARLOS F
bstantia ALVAREZ CA
now and ANDEZ IEA
y in para OF CONTEN
other keyIVE SUMMAR
n strateg UPPLY TRA
lected sce COAL MET
ffer substaCOAL PRI
with respecTS PRICE
ir reliancAL MINING
on captureE ORT GE

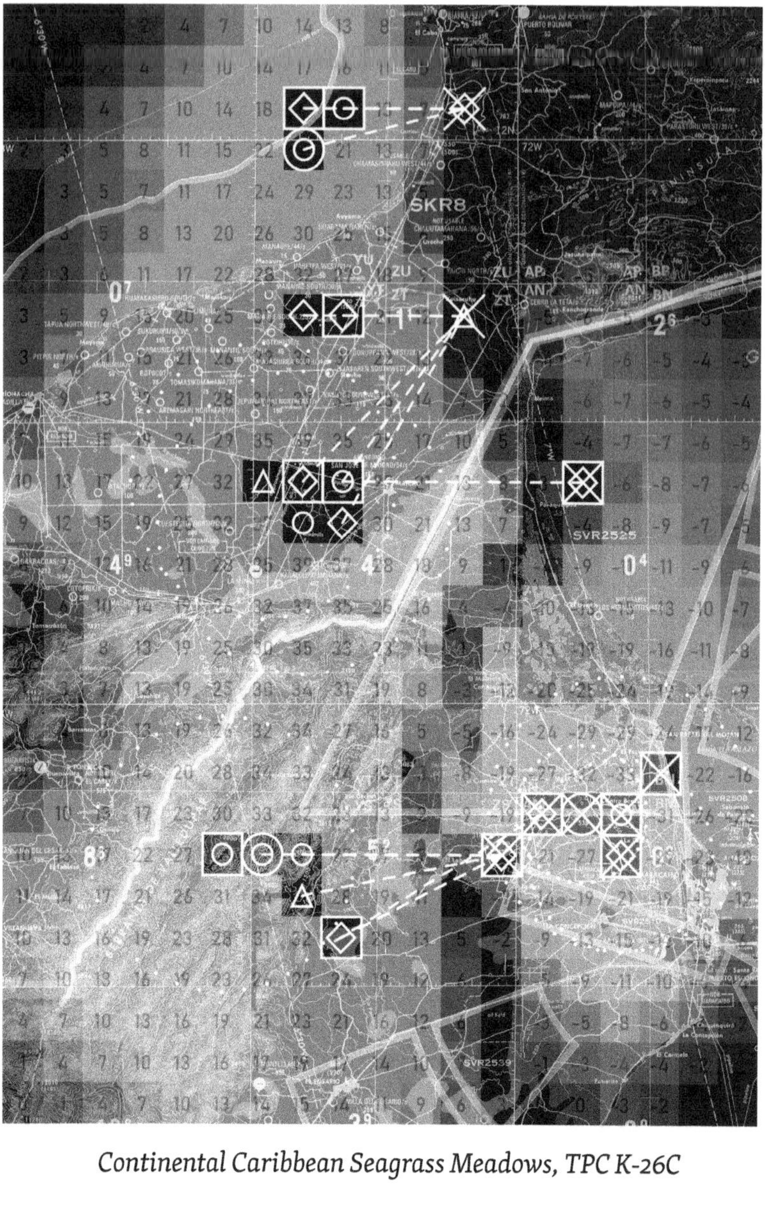

Continental Caribbean Seagrass Meadows, TPC K-26C

euracoal NON
hem schiff AND
aul simons yNOT W
versity and CTI
abumoto j p T THE
individualEVASTAT
ganisations A MULT
ntributed toITICAL
eport are no CAUSES
nsible for anTHERE
ion or judgem D BY
contains any CTIVE
or omission VED A
sole responsBY A B
y of the iea AND AM
estions and S OF PE
s please coP MYSELF
rlos fern nMERA COO
rez carlos A I USE
z iea org tM MINERA
contents exLASTICS
summary demaS SOME
ly trade theE FORME
al metallurgST OF T
al prices andIONS O
prices costsND EXT
ining projec AT

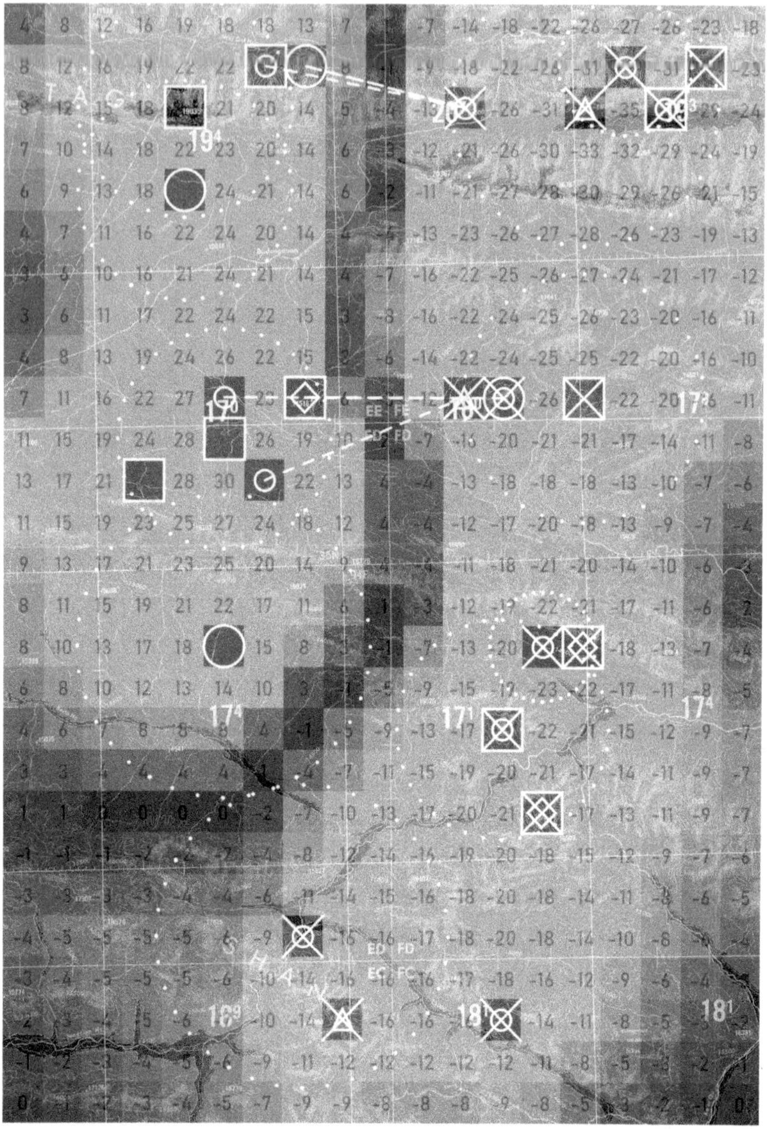

Juniperus Komarovii Forest, TPC G-8D

ies in the THE TOP
ast northFOR THE MI
gion contLS SECTOR
credit k IS LITTLE
ukiafp viION THAT E
ages the S WILL DIM
ange rela INVESTORS
and projS IN MININ
tlined abLIFTING TH
t seriousUNCERTAINT
ons for agAHEAD THE
and food TALS SECT
and for th STAKEHOL
environmeTO BE REA
six countTHING IN
ined in tOR NOT REB
especiaPBELL DAMIE
nsidereAND JOHN TIV
backgroORE L OOKING
p PROVED REMARK
ABLY TUMULTUOUS FOR
COMMODITIES MARKET
S WILL OFFER MORE C
ERTAINTY TO GET AN
IDEA WHITE CASE HAS
CONDUCTED ITS SEVE
TH ANNUAL SURVEY

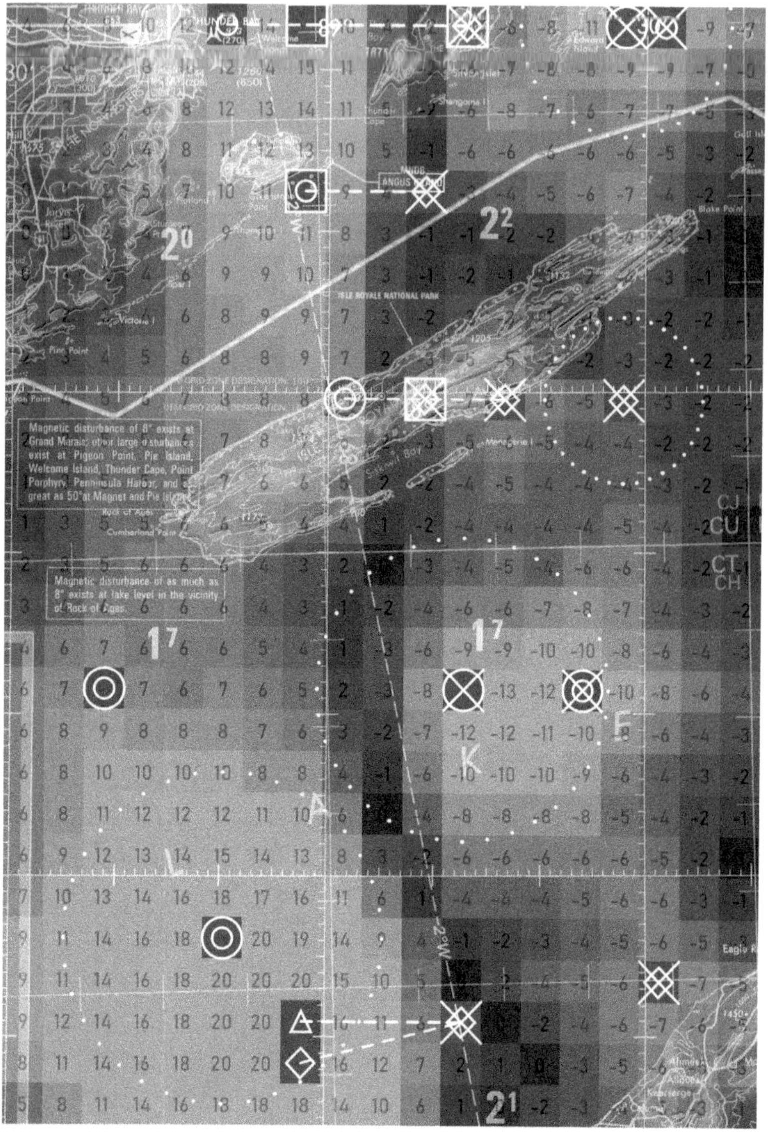

Larix gmelinii, Pinus sylvestris var. mongolica forest, TPC G-9D

y prices CAL PLEDG
ve pre co SECTOR RE
 in many NOMY NO PR
wholesal MEXICO NO
ty pricENT NO DATA
ountrie IN LAW NO D
om theNORWAY NO COM
 obseNTA NO DATA QA
ook NO COMMITMENT N
 wiRODUCTION NO DAT
ric RUSSIAN FEDERAT
mmodIN STRATEGY DOC
aturT SAUDI ARABIA
 theTICAL PLEDGE N
  g UCTION SOUTH A
    A IN STRATEGY D
   MENT NO DATA NO
pe   O DATA UAE ND
age by TIVE NO PROD
 from re B UK IN LA
 in despi US IN POL
ices in euNT A NORW
 still rougITTED TO
le levels wSION SOC
s prices inITS CLIM
ly about hiACT WITH
n in uncertEDUCTION

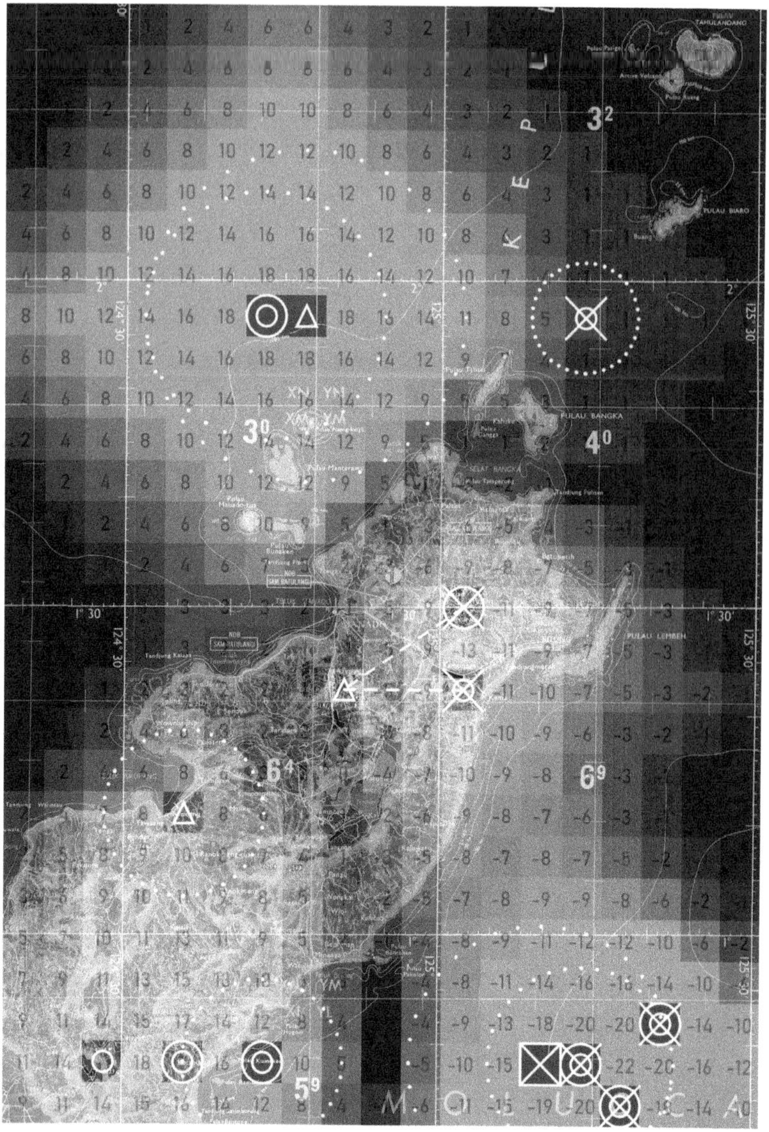

Mangroves of the Western Coral Triangle, TPC L-12D

h was pri
pported by
orth america
he gas rich mar
in africa and t
middle east china
egained its positi
n as the world s la
rgest lng importer
with natural gas
mand increasing
in contrastO O
gas consu GOVER
urope fS THERE
loweAGMENTED E
secuF INCREASI
y concE SECTOR AD
bility INTERVENT
emission VER THE
efforts N MOST S
g the dirWATER FOO
gas relaCULTURE TR
es the glND INFRAST
y crisis URISM THE
by russia INTERNAL
n of ukra TS BY LAR
curity ofIES FINANC

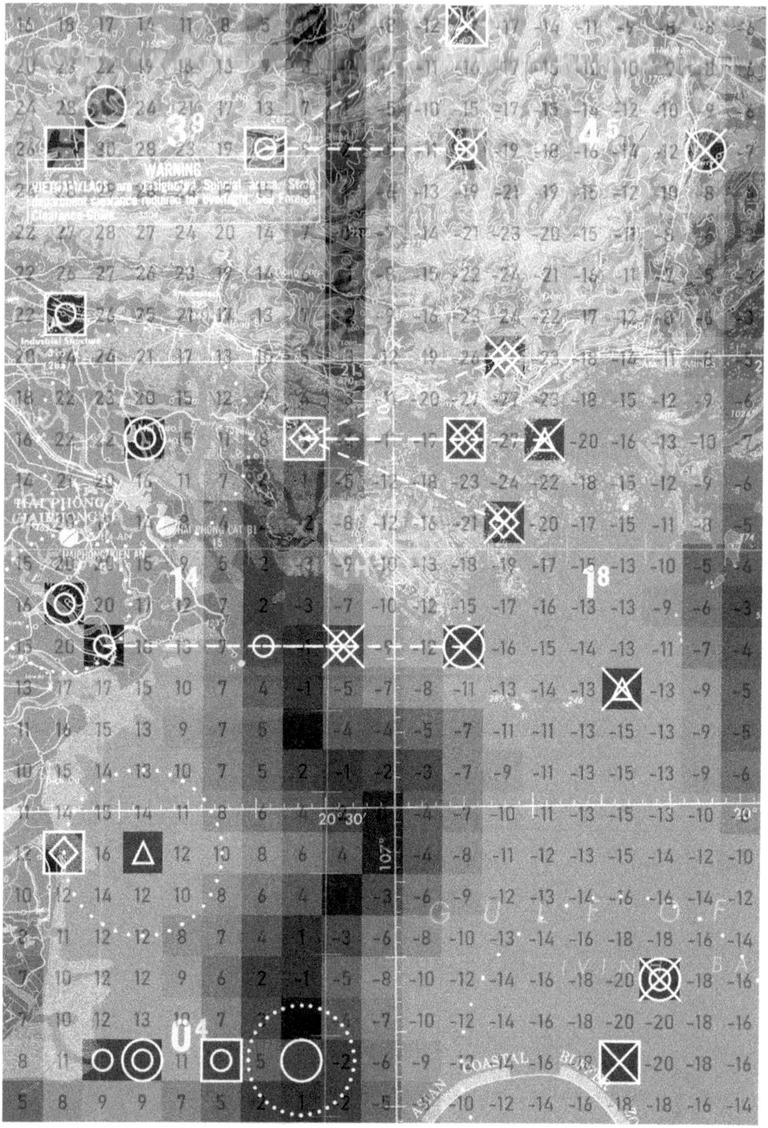

Mangroves of the South China Sea, TPC J-11A

using o OCH STRATIG
the infAND GEOLOGI
the momrONSIDERING
expresseANTHROPOGE
omr are tTS INCLUDI
e opec s RTAINLY NO
and do nD TO THE RI
ily reflRICUL TURE
ws of itNDANT DEFOR
bodies THE EXTRAC
ountriesCOAL OIL AN
nation oD THEIR ATM
cal entC CONSEQUENC
momr aCOMBUSTION O
nd presN BASED FUEL
data anMISSIONS COR
rials d LOSS OCEAN
the exp ATION SOIL
ny opiniION A RATE
er on the RM EXTINC
pec and o RING AT T
er count OF TIMES H
ning theAN THROUGHO
us of a OF THE LAST
erritor LLION YEARS
of its HAPS MOST S
or concGLY A RATE

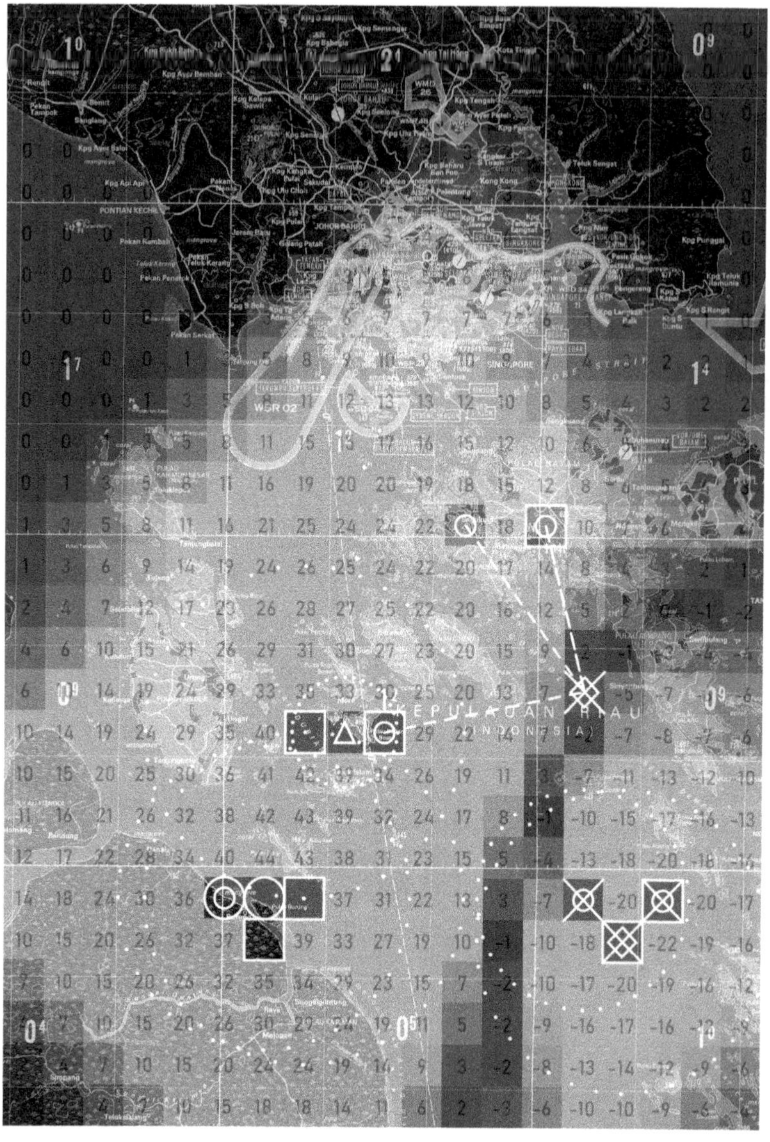

Mangroves of the Sunda Shelf, TPC L-10C

n
g s
nergy
policie
lt global
sumption in
to be lower
in although chi
will have the la
word coal s shif
to asia is accele
rating the dominanc
e of china in coal
markets is stron AN
than any other ON
ntry for any o INTE
fuel it consumTHE L
re than half oTIGAT
world s coalUNNECE
roduces half NTION
and it is theAKES P
st importer CONTEXT
ing for closAD ECOL
e third of tPSE AND
al coal tradINVESTE
dia and aseRING WH

Sources

War Machine was composed on a largely ad-hoc basis from September 2021 until June 2024. Over this time, it employed a mass of found and remixed materials derived from a variety of online sources, chiefly extracts from varied published marketing reports concerning the global state of oil, gas, and mining, as well as CC-licensed academic texts. Therefore, the sources below should only be taken as an illustrative sample of some of the materials used in the composition of each plate.

The Tactical Pilotage Charts for all the regional maps were sourced from the open access Perry-Castañeda Library Map Collection at the University of Texas at Austin. They are available online at the now archived site https://maps.lib.utexas.edu/maps/tpc/.

An illustrative sample of the textual sources employed for *War Machine* include:

Beuchle, R., F. Achard, C. Bourgoin, and C. Vancutsem. 2021. *Deforestation and Forest Degradation in the Amazon: Updated Status and Trends for the Year 2021*. Publications Office of the European Union. https://publications.jrc.ec.europa.eu/repository/handle/JRC130081.

Climate Assembly UK. 2020. *The Path to Net Zero: Climate Assembly UK Report*. https://www.climateassembly.uk/recommendations/index.html.

Davis, Heather, and Etienne Turpin, eds. 2014. *Art in the Anthropocene: Encounters Among Aesthetics, Politics, Environments and Epistemologies*. Open Humanities Press.
International Energy Agency (IEA). 2023. *Coal 2023*. IEA. https://www.iea.org/reports/coal-2023.
———. 2023. *Critical Minerals Market Review 2023*. IEA. https://www.iea.org/reports/critical-minerals-market-review-2023.
———. 2024. *Electricity 2024*. IEA. https://www.iea.org/reports/electricity-2024.
———. 2024. *Gas Market Report, Q1 2024*. IEA. https://www.iea.org/reports/gas-market-report-q1-2024.
Litvintseva, Sasha. 2023. *Geological Filmmaking*. Open Humanities Press.
Miller, Kathryn A., George Zittis, David Santillo, and Paul Johnston. 2022. *Living on the Edge: The Implications of Climate Change for Six Countries in the Middle East North Africa Region*. Greenpeace Research Laboratories Technical Report. https://www.greenpeace.to/greenpeace/?p=4611.
Office of the Director of National Intelligence. 2023. *2023 Annual Threat Assessment of the U.S. Intelligence Community*. ODNI. https://www.dni.gov/index.php/newsroom/reports-publications/reports-publications-2023/3676-2023-annual-threat-assessment-of-the-u-s-intelligence-community.
Stockholm Environment Institute (SEI), International Institute for Sustainable Development (IISD), Overseas Development Institute (ODI), E3G, and UNEP. 2023. *Production Gap Report* 2023. UNEP. https://www.unep.org/resources/production-gap-report-2023.

United Nations Environment Programme (UNEP). 2023. *Climate Adaptation Gap Report 2023*. UNEP. https://www.unep.org/resources/adaptation-gap-report-2023.
Wakefield, Colin. 2023. *Anthropocene Back Loop*. Open Humanities Press.

www.ingramcontent.com/pod-product-compliance
Lightning Source LLC
LaVergne TN
LVHW020235210926
848481LV00016B/2212
* 9 7 8 1 6 8 5 7 1 3 2 1 8 *